HEALTHCARE TECHNOLOGIES SERIES 56

Affective Computing Applications Using Artificial Intelligence in Healthcare

IET Book Series on e-Health Technologies

Book Series Editor: Professor Joel J.P.C. Rodrigues, College of Computer Science and Technology, China University of Petroleum (East China), Qingdao, China; Senac Faculty of Ceará, Fortaleza-CE, Brazil and Instituto de Telecomunicações, Portugal

Book Series Advisor: Professor Pranjal Chandra, School of Biochemical Engineering, Indian Institute of Technology (BHU), Varanasi, India

While the demographic shifts in populations display significant socio-economic challenges, they trigger opportunities for innovations in e-Health, m-Health, precision and personalized medicine, robotics, sensing, the Internet of things, cloud computing, big data, software-defined networks, and network function virtualization. Their integration is however associated with many technological, ethical, legal, social, and security issues. This book series aims to disseminate recent advances for e-Health technologies to improve healthcare and people's wellbeing.

Could you be our next author?

Topics considered include intelligent e-Health systems, electronic health records, ICT-enabled personal health systems, mobile and cloud computing for e-Health, health monitoring, precision and personalized health, robotics for e-Health, security and privacy in e-Health, ambient assisted living, telemedicine, big data and IoT for e-Health, and more.

Proposals for coherently integrated international multi-authored edited or co-authored handbooks and research monographs will be considered for this book series. Each proposal will be reviewed by the book Series Editor with additional external reviews from independent reviewers.

To download our proposal form or find out more information about publishing with us, please visit https://www.theiet.org/publishing/publishing-with-iet-books/.

Please email your completed book proposal for the IET Book Series on e-Health Technologies to: Amber Thomas at athomas@theiet.org or author_support@theiet.org.

 The Institution of Engineering and Technology

Affective Computing Applications Using Artificial Intelligence in Healthcare

Methods, approaches and challenges in system design

Edited by
M. Murugappan

The Institution of Engineering and Technology

Published by The Institution of Engineering and Technology, London, United Kingdom

The Institution of Engineering and Technology is registered as a Charity in England & Wales (no. 211014) and Scotland (no. SC038698).

The Institution of Engineering and Technology
Futures Place
Kings Way, Stevenage
Hertfordshire, SG1 2UA, United Kingdom

www.theiet.org

British Library Cataloguing in Publication Data
A catalogue record for this product is available from the British Library

ISBN 978-1-83953-731-8 (hardback)
ISBN 978-1-83953-732-5 (PDF)

Typeset in India by MPS Limited

Cover Image: janiecbros/E+ via Getty Images

To my wife Subbu, my children Sakthi, Harish, and my parents

Contents

About the editor

Professor Dr. M. Murugappan has been working at Kuwait College of Science and Technology (KCST), Kuwait as a Full Professor in Electronics, Department of Electronics and Communication Engineering since 2016. It is also his pleasure to serve as a Visiting Professor at the School of Engineering at Vels Institute of Science, Technology, and Advanced Studies in India since 2022. He is also an International Visiting Fellow at the Center of Excellence in Unmanned Aerial Systems at Universiti Malaysia Perlis in Malaysia. In 2006, he graduated from Anna University, India with an M.E. degree in Applied Electronics. He received his Ph.D. from Universiti Malaysia Perlis, Malaysia in 2010 for his contribution to the field of Mechatronic Engineering. Between 2010 and 2016, he worked as a Senior Lecturer at the School of Mechatronics Engineering, Universiti Malaysia Perlis, Malaysia. In this role, he taught a variety of courses related to biomedical and mechatronics engineering. As a result of his excellent publications and research products, he has been awarded several research awards, medals, and certificates. In a study by Stanford University, he was recently ranked among the top 2% of scientists working in experimental psychology and artificial intelligence for three consecutive years (2020–2022). His research into affective computing has received more than 750K in grants from Malaysia, Kuwait, and the UK. His publications include more than 140 peer-reviewed conference proceedings papers, journal articles, and book chapters. Several of his journal articles have been recognized as best papers, best papers of the fiscal year, etc. Prof. Murugappan is a member of the editorial boards of PLoS ONE, Human Centric Information Sciences, Journal of Medical Imaging and Health Informatics, and International Journal of Cognitive Informatics. In addition to being the Chair of the IEEE Kuwait Section's Educational Activities Committee, he is an active reviewer for IEEE Transactions on Multimedia, IEEE Transactions on Affective Computing, IEEE Transactions on Health Informatics, and IEEE Transactions on Biomedical Signal Processing. He is interested in affective computing, affective neuroscience, the Internet of Things (IoT), the Internet of Medical Things (IoMT), cognitive neuroscience, brain–computer interface, neuromarketing, neuroeconomics, medical image processing, machine learning, and artificial intelligence. Throughout his career, he had a strong collaborative relationship with leading researchers in the fields of biosignal/image processing, artificial intelligence, the Internet of Things, and cognitive neuroscience. He is a member of several international professional societies including IEEE, IET, IACSIT, IAENG, and IEI.

Preface

Over the past few years, affective computing has progressed tremendously and has become widely accepted in a wide variety of applications, such as e-learning, marketing, customer service, smart transportation, healthcare, and many others. Different modalities, such as facial expressions, biosignals, gestures, and speech, are used in affective computing to detect different emotions (happy, angry, sad, fearful, disgusting, and surprised) of people. A growing number of healthcare facilities are utilizing this technology for clinical diagnosis in recent years. Additionally, an automated diagnosis system using affective computing has been developed to identify emotional disturbances or impairments in patients. In the past, it has been primarily used as a tool for diagnosing emotional disturbances associated with a variety of diseases, such as depression, stress, emotional stress, anxiety, bipolar disorder, attention deficit disorder, strokes, Parkinson's disease (PD), autism spectrum disorders (ASD), mood disorders, etc. In most emotional computing applications, facial expressions and physiological signals are used to assess emotional state changes in subjects, such as electrocardiograms (ECGs), electromyograms (EMGs), electroencephalograms (EEGs), electrooculograms (EoGs), skin temperature (ST), galvanic skin response (GSR), photoplethsymograms (PPGs), etc. Researchers have mostly used biosignals and facial images to extract meaningful information about different emotions. Machine learning algorithms and deep learning algorithms have been used to make decisions using artificial intelligence (AI).

This book is mainly intended to discuss some of the state-of-the-art research works on affective computing applications in healthcare using multi-modal inputs. Researchers from different countries have contributed to developing this book, *Affective Computing Applications Using Artificial Intelligence in Healthcare*. There were nine chapters selected for publication after being reviewed by 31 reviewers anonymously from 25 chapters. There were several topics covered in the book, including hybrid deep neural network (DNN)-based emotion recognition from facial and thermal images, biosignals-based gender and emotion recognition through AI techniques, gestures-driven rehabilitation system for post-stroke patients using affective computing methods, a deep neural network for depression detection from emotional speech signals, pre-trained DNN-based emotion recognition from EEG signals, a novel stress detection method based on biosignals and deep learning algorithms, explainable AI models based emotion recognition using AI, and the convergence of emotion recognition using the Internet of Artificial Intelligence Things (IoAIT) in the healthcare industry and medical domain.

As the book's editor, I would like to thank the authors for revising and improving their papers based on feedback from our reviewers. Also, I would like to thank the reviewers for setting up a time to review the papers within their schedules. As a result, they will be able to provide constructive feedback on their content and suggestions for improvement. It is a pleasure for me to serve as editor of this book. With sincere gratitude, I acknowledge the authors and reviewers who have contributed to the creation of this book.

M. Murugappan
Editor

Acknowledgement to reviewers

S. No.	Name and affiliation
1	Dr. John Thomas, McGill University, Montreal, Canada
2	Dr. Prabha Sundaravadivel, University of Texas at Teyler, USA
3	Dr. U. Raghavendra, Manipal Institute of Technology, Karnataka, India
4	Dr. Usman Akram, National University of Sciences, Pakistan
5	Dr. Jerritta Selvaraj, Vels University, Chennai, India
6	Dr. Muhammad Enamul Hoque Chowdhury, Qatar University, Qatar
7	Dr. Ali K. Bourisly, Department of Physiology, Kuwait University, Kuwait
8	Dr. N. Punitha, SSN College of Engineering, Chennai, India
9	Dr. N. P Guhan Seshadri, National Institute of Technology Raipur, India
10	Dr. Bikesh Kumar Singh, National Institute of Technology Raipur, India
11	Dr. Wan Khairunizam, Universiti Malaysia Perlis (UniMAP), Malaysia
12	Dr. Saidatul Adreenawati Awang, Universiti Malaysia Perlis, Malaysia
13	Dr. Anandhi, Agni College of Engineering, Chennai, India
14	Dr. Sumithra Palanisamy, Dr. N.G.P. College of Technology, India
15	Dr. Ramanathan Subramanian, University of Canberra, Australia
16	Dr. Mufti Mahmud, Nottingham Trent University, United Kingdom
17	Dr. Suma Dawn, Jaypee Institute of Information Technology, India
18	Dr. R. Balasubramanian, NMAM Institute of Technology, India
19	Ms. Muthumeenakshi Subramanian, Case Western University, USA
20	Dr. Hanif Heidari, Damghan University, Iran
21	Dr. Amalin Prince, Birla Institute of Technology and Sciences, India
22	Dr. Yuvaraj Rajamanikam, Nanyang Technological University, Singapore
23	Dr. P. Karthikeyan, Madras Institute of Technology, Anna University, India
24	Dr. Rodrigo S. Jamisola, Botswana International University of Science and Technology, Botswana
25	Dr. Ateke Goshvarpour, Sahand University of Technology, Iran
26	Dr. Shamin Kaisar, Jahangirnagar University, Bangladesh
27	Dr. Mohamed Yacin Sikkandar, Majmah University, Saudi Arabia
28	Dr. Sneghalatha, SRM Institute of Science and Technology, India
29	Dr. B. Geethanjali, SSN College of Engineering, India
30	Dr. Raja Lakshmi, Vellore Institute of Technology, India
31	Dr. Gowri Annasamy, Indian Institute of Information Technology, Design and Manufacturing, Kanchipuram, India

Chapter 1

EEG-based emotion recognition using time–frequency images and hybrid ResNet models

Shreyas Krishnan[1], Lavanya Murugan[2], Aya Hassouneh[3], Rajamanickam Yuvaraj[4], A. Amalin Prince[5], T. Thiyagasundaram[6] and M. Murugappan[7,8]

Emotion recognition involves identifying and interpreting human emotions based on inputs such as facial expressions, speech, and physiological signals. Electroencephalography (EEG), which measures the electrical activity of the brain, is one of the physiological signals that can be used to determine emotion. Many machine learning techniques have been used to tackle the challenging task of identifying emotions from EEG waves. By using different machine learning techniques, this study addresses the problem of emotion recognition and classification into valence and arousal and compares crucial metrics such as accuracy, precision, recall, and F1-score. The aim of this study was to combine various machine learning models with ResNet's powerful feature extraction capability to improve emotion recognition accuracy using two datasets: Database for Emotion Analysis using Physiological Signals (DEAP) and multimodal database consisting of electroencephalogram (EEG) and electrocardiogram (ECG) signals (DREAMER). There were three ResNet models compared in the study: ResNet152-XGBoost, ResNet152-Extreme Learning Machines (ResNet152-ELM), and ResNet-152. In our experiments, the hybrid models outperformed the pure ResNet model, with ResNet-ELM having the highest mean accuracy of 99.34% and 92.18% on both DREAMER

[1]Department of Electrical and Electronics Engineering, BITS Pilani Goa Campus, India
[2]School of Engineering and Computing, American International University, Kuwait
[3]Electrical and Computer Engineering Department, Western Michigan University, USA
[4]Science of Learning in Education Centre (SoLEC), Office of Education Research, National Institute of Education, Nanyang Technological University, Singapore
[5]Department of Electrical and Electronics Engineering, BITS Pilani Goa Campus, India
[6]Department of NoiNadal, Sri Sairam Siddha Medical College and Research Centre, West Tambaram, India
[7]Intelligent Signal Processing (ISP) Research Lab, Department of Electronics and Communication Engineering, Kuwait College of Science and Technology, Kuwait
[8]Department of Electronics and Communication Engineering, Faculty of Engineering, Vels Institute of Sciences, Technology, and Advanced Studies, India

and DEAP datasets. The results demonstrate that hybrid ResNet models can accurately identify human emotions from EEG data and also with the state-of-the-art methods reported in the literature.

Keywords: Emotion recognition; EEG signals; ResNet; Machine learning; Hybrid models

1.1   Introduction

According to Hockenbury *et al.*, emotions are complex psychological states that involve three distinct components: a subjective experience, a physiological response, and a behavioral or expressive response [1]. People's emotions have a profound effect on many aspects of their lives, including memorization, learning development, communication with others, and many others. Researchers are researching and developing new techniques for recognizing emotions in the field of affective computing [2,3]. In accordance with Ekman *et al.*, fundamental emotions have a biological basis and function adaptively to facilitate how people cope and respond to their environment [4]. Picard *et al.* define emotion recognition as the ability to identify and interpret emotions across a range of sources, including facial expressions, voice, and physiological signals [5]. They propose a method for collecting and assessing physiological data, such as heart rate, skin conductance, and facial expression recognition, to reveal emotional information about an individual. Facial expressions and speech were used as metrics in early emotion detection systems proposed by Song *et al.* and Essa *et al.* [6,7]. Due to the potential for intentional variation in facial expressions and speech, these metrics resulted in some false positives [8]. An audio-visual induction protocol that uses *db4* wavelet in the frequency band to improve K-nearest neighbors (KNN) based classification of electroencephalogram (EEG) signal was proposed in [9,10] and is used for classifying emotions into six basic types (happiness, anger, sadness, fear, disgust, and surprise) using EEG signals. The two-dimensional emotion model is illustrated in Figure 1.1. Every emotion

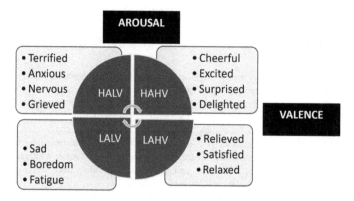

Figure 1.1   Two-dimensional emotion model

is represented as some combination of arousal and valence, which are classified independently as HIGH/LOW.

The method we developed combines the strengths of ResNet and machine learning models to improve the accuracy of emotion recognition from EEG signals. The following contributions are made by our study:

- Combining ResNet and machine learning models, we propose a novel method for recognizing emotions using EEG signals.
- We evaluate the performance of four hybrid ResNet models to recognize emotions based on EEG data.
- We demonstrate that our hybrid ResNet models perform better than pure ResNet models in recognizing emotions.

The rest of the paper is structured as follows: Section 1.2 provides a detailed description of our proposed methodology, including data acquisition and preprocessing, feature extraction, and the hybrid ResNet models used in this study. The results of our experiment are presented in Section 1.4 along with a comparison of the hybrid ResNet model performance compared to the pure ResNet model performance. In Section 1.5, we include the limitations and provide future work in Section 1.6. Finally, Section 1.7 concludes the research findings and conurbations.

1.2   Literature review

Electroencephalography (EEG) is one such physiological signal that examines brain electrical activity non-invasively. Identifying emotions from EEG signals requires several steps to satisfy brain–computer interface requirements [9,10]. The four basic steps of emotion recognition are signal acquisition, signal preprocessing, identifying features in EEG signals, and classifying emotions [11]. EEG signals are classified according to the different frequency ranges as Delta (δ) waves with a frequency between 0 and 4 Hz, Theta (θ) waves with a frequency range of 4–7 Hz, Alpha (α) waves with a frequency range of 8–13 Hz, Beta (β) wave with the frequency range of 13–30 Hz, and Gamma (γ) with frequencies above 30 Hz. A strong correlation exists between the different EEG waves and the emotional activities of the brain [12].

During the preprocessing phase, various unwanted interferences are removed such as electrooculogram artifacts, power–frequency interference, electromagnetic interference, etc. The preprocessing of signals is carried out by a variety of techniques, including principal component analysis, blind source separation, detrended fluctuation analysis, and notch filters [11]. Most often, two or more techniques are combined to achieve accurate emotion recognition [13]. In recent years, feature extraction methods such as adaptive autoregression, fast Fourier transform (FFT), wavelet transform (WT), and common spatial pattern have been used with parameters from time, frequency, and spatial information [14,15].

Using machine learning techniques, Koelstra *et al.* classified emotions based on physiological signals and annotated the data with self-reported emotion ratings [16]. Research in machine learning has shown great promise for detecting real-time emotions from EEG signals. A variety of machine learning techniques have been successfully employed for this purpose, including support vector machines (SVMs), random forests, and deep learning (DL) [16]. In artificial intelligence, DL has revolutionized numerous domains, especially image recognition [17]. A number of algorithms are used to detect deep features, such as long short-term memory (LSTM) [18] and recurrent neural network (RNN) [19]. For emotion recognition, Craik *et al.* [20] demonstrate the dominance of deep neural networks by using deep belief networks (DBNs), RNNs, and convolutional neural networks (CNNs).

Expanding on this research, more investigations advance our understanding of how EEG signals can be used to identify emotions using EEG signals. Yuvaraj *et al.* proposed a novel approach by employing a three-dimensional CNN with ensemble learning techniques for emotion recognition from spatio-temporal EEG signal representations [21]. Maithri *et al.* [22] explored current and upcoming trends in automated emotion recognition, highlighting the dynamic nature of this developing field. Hassouneh *et al.* [23] developed a real-time emotion recognition system using facial expressions and EEG, employing machine learning and deep neural network methods.

Human–computer interfaces are becoming more capable of accurately recognizing human emotions. Several methods have been proposed for the classification of emotions based on EEG signals to date [17,19]. In Song *et al.* [24], nonlinear dynamical features were extracted using power spectral density and SVM. In 2011, Nie *et al.* [25] proposed a series of bandpass filters followed by FFT. A short-time Fourier transform (STFT) with a non-overlapping Hanning window was employed by Lin and Liu *et al.* [26] for extracting features, and SVM and linear discriminant analysis (LDA) were applied for classification. Lee and Lee [27] used STFT with CNN. According to Murugappan [28], four different wavelets were used in wavelet analysis and feature extraction. Using statistical parameters extracted by Veropoulos *et al.* [29], a higher-order crossing was analyzed using LDA, KNN, and SVM. Guo *et al.* [30] proposed using hybrid SVM where features were extracted by WT and fuzzy cognitive maps. Bajaj *et al.* [31] used multiwavelet (MWT) analysis with three MWTs. There are two forms of MWTs: (i) Orthogonal type, like Gernoimo-Hardin-Massopust Chui Lian and Symmetric asymmetric (SA4), and (ii) Bi-orthogonal, including Bi-Orthogonal Hermite (Bih52). Three-level MWTs are used for emotion classification.

Over the past few decades, a diverse array of DL methods and models, as well as their combinations, have been used for emotion detection. The feature classifications in this context are performed using a multiclass least square SVM. In 2014, Zheng *et al.* [32] proposed a method for identifying emotions based on DBNs and hidden Markov models. Based on the Hilbert–Huang spectrum, Zhao–Atlas–Marks distribution, and spectrogram methods, Hadjidimitriou and Hadjileontiadis [33] developed a hybrid emotion classification method using SVM and KNN. A recent study by Yuvaraj *et al.* [34] investigated the features that could be used to recognize emotions. Using five public datasets, the study identified the most significant EEG features for different emotional states classification.

The ResNet architecture has shown exceptional performance in recognizing images, detecting objects, and segmenting images [35]. The innovation of ResNet lies in the use of skip connections, which bypass certain layers and allow information to flow directly to subsequent layers, overcoming the degradation problem faced by deep neural networks [35]. Therefore, ResNet can effectively train deep neural networks with hundreds or even thousands of layers [36,37]. Besides studies involving EEG and emotion analysis, ResNet has also been found in other fields [38]. Using a DL model based on the ResNet-50 architecture, ResNet has been used in EEG analysis to classify brain signals according to different environmental stimuli [39]. A discussion of CNN with ResNet50, SEED dataset, DREAMER, and AMIGOS is presented in Topic and Russo [40]. EEG signals were used to recognize emotions using different methods with varying degrees of accuracy, ranging from 50% to 95%.

In this study, we combine ResNet with various machine learning models to enhance the precision of EEG emotion recognition. Our specific focus involves a comparative evaluation of ResNet152-XGBoost, ResNet152-ELM, and ResNet-152 models. Besides accuracy (ACC), we also compare the precision (PRE), recall (REC), and F1-score (F1-S) metrics for the two different datasets DEAP and DREAMER. This study aims to explore the potential for detecting emotions precisely using EEG signals based on hybrid ResNet models.

1.3 Materials and methods

An overview of the research methodology can be found in Figure 1.2. This diagram illustrates how we independently classified emotions as HIGH or LOW using our representation of arousal and valence as a combination, resulting in two distinct binary classification problems. The ResNet model was used for feature extraction following data preprocessing and time–frequency image analysis. In the end, we employed machine learning models to perform binary classification, where outputs were either high or low depending on the model.

1.3.1 Dataset

In our study, we used two different datasets, DEAP and DREAMER.

A DEAP emotion dataset, which includes both peripheral nervous system (PNS) signals and 32-channel EEG recordings, was developed by Koelstra *et al.* [16]. The following are the specifications of the dataset.

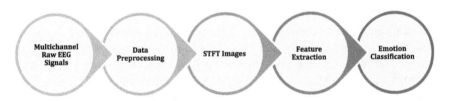

Figure 1.2   Modules of our emotion recognition system

- In this dataset, 32 healthy participants watched 40 music videos, resulting in 40 trials in total.
- A 1-min music video was used in each trial, followed by self-assessment ratings of arousal, valence, dominance, like/dislike, and familiarity.
- There are 3 s of baseline data in each video, as well as 60 s of EEG recordings at a sampling frequency (Fs) of 512 Hz.

Katsigiannis and Ramzan [41] describe the DREAMER dataset as consisting of ECG signals recorded on two channels and EEG recordings on 14 channels with the following features:

- Twenty-three healthy volunteers, aged 22–33, watched 18 video clips varying in duration from 65 to 393 s while these signals were recorded.
- Using a self-assessment manikin, participants evaluated their levels of arousal, valence, and dominance after each clip.
- A baseline signal of 60 s was also recorded before each video clip.
- An emotive EPOC wireless neuro headset with a sampling rate of 128 Hz was used to record the EEG signals.

1.3.2   Data preprocessing

EEG signal preprocessing, feature extraction, and emotional state categorization were performed using Python (v3.7.1) and MATLAB (vR2020b). On average, 847 EEG trials were conducted across datasets (standard deviation = 612.4), 1280 for DEAP (40 trials with 32 participants), and 414 for DREAMER (18 trials with 23 participants).

Data from EEG trials were filtered with 50/60 Hz notch filters and fourth-order Butterworth filters to remove electrical mains and DC artifacts. We down-sampled the data to 128 Hz to match the sample rates of the different datasets, segmented it into 1-s non-overlapping intervals, and then re-referenced it to the average of all the intervals. Then, epochs were treated to an automatic artifact rejection system to eliminate eye blinks and other electrical artifacts by removing segments with data over 100 μV. The segmented epoch was then input to STFT. The participants who were accepted for further analysis had an average of 1046 (SD = 411) valence epochs and 1036 (SD = 438) arousal epochs. In Figure 1.3, we present time–frequency analysis, or STFT, examples of our model. A Hamming window of a 1-s length was applied to each channel's signal to prepare the EEG data for classification. After vertically concatenating the images, a single input image is created for each trial.

1.3.3   Proposed deep CNN architecture

Our study proposes a deep learning approach for classifying EEG signals by using time–frequency images. STFTs were used to convert raw EEG signals into time–frequency representations, which were then transformed into images using spectrograms. Essentially, each EEG signal is converted into an STFT image and concatenated on top of each other to form an input image. The models we used in this study included XGBoost due to its effectiveness in handling large-scale

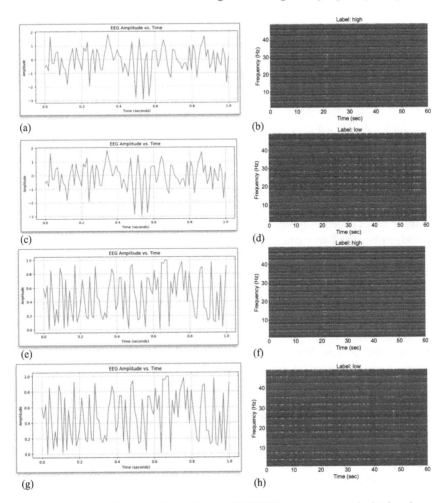

Figure 1.3   The 1s-duration low-valence: (a) EEG segment example-high valence;
(b) STFT magnitude of high valence; the 1s-duration low-arousal; (c)
EEG segment example-low valence; (d) STFT magnitude of low
valence; the 1s-duration high-valence; (e) EEG segment example-low
arousal; (f) STFT magnitude of high arousal; (g) EEG segment
example; (h) STFT magnitude of low arousal; and the 1s-duration
high-arousal

datasets [38], SVM chosen for its effectiveness in separating data with clear
boundaries [42], extreme learning machines (ELM) valued for its simplicity and
efficiency [43], and LSTM favored for its effectiveness in handling sequential data
[44]. These models were used to categorize labels associated with valence and
arousal. Extracted features from the ResNet-152 (shown in Figure 1.4) model were
used to develop these models.

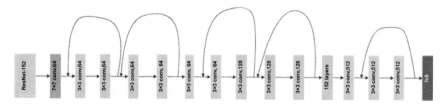

Figure 1.4   Graphical representation of configured ResNet-152-based proposed CNN architecture

A grid search was performed for each hyperparameter to determine the optimal set of parameters. Automated hyperparameter tuning techniques, such as Bayesian optimization or genetic algorithms, can also be used to find the best hyperparameter set and explore the hyperparameter space efficiently.

In order to assess the reliability and generalizability of our proposed model, we used two distinct cross-validation (CV) techniques, incorporating subject-correlated inference methods (SCR). The first method is a five-fold CV, which is designed specifically for subject-correlated detection. By using this method, a dataset is divided into five equal subsets, four of which are used for training the model, and the fifth is used for testing it. Every subset is tested exactly once, and this process is repeated five times. In this way, the model can detect patterns that are consistent across subjects, including those that are specific to individual subjects.

As for the second method, we used leave-one-subject-out (LOSO) CV, which is suitable for subject-independent detections. Using this technique, one subject's data is removed from the training set, and the model is trained with the remaining data. Data from the excluded subjects is then used as the testing set, and the same process is repeated for all subjects. Using this method, we can test how well the model performs on data from unseen subjects, which is crucial for generalizing. Figure 1.5 shows a concise summary of the model architecture. Detailed parameters for each layer of our proposed ResNet-152 model are shown in Table 1.1.

1.3.4   Evaluation metrics

A dataset is divided into subsets, and each subset is alternately used as the testing set while the remaining subsets collectively form the training set. The model undergoes training on the training set and is subsequently evaluated on the testing set. The process is repeated k times, with each fold being used once as a testing set. Afterward, each fold's performance metrics are averaged to estimate the model's overall performance. A model's performance is measured by its accuracy (ACC) (1.1), recall (REC) (1.2), F1-score (F1-S) (1.3), and precision (PRE) (1.4).

The formulas for the mentioned metrics are as follows:

$$ACC = \frac{(TP + TN)}{(TP + TN + FP + FN)} \tag{1.1}$$

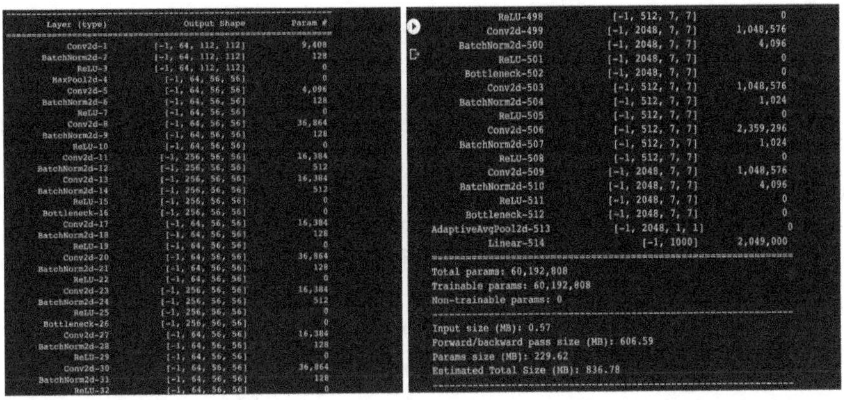

Figure 1.5 *The proposed model architecture shows the different parameters used in this study*

Table 1.1 *Details of parameters of each layer in the configured ResNet-152*

Layer	Layer name	No. of filter	Filter length	Stride	Output shape	Parameters
0	input	–	–	–	[3, 224, 224]	–
1	Conv2D-1	64	(7, 7)	(2, 2)	[−1, 64, 112, 112]	9408
2	BatchNorm2d-2	64	(1, 1)	(1, 1)	[−1, 64, 112, 112]	128
3	ReLU-3	64	(3, 3)	(1, 1)	[−1, 64, 112, 112]	0
4	MaxPooling-4	256	(1, 1)	(1, 1)	[−1, 64, 56, 56]	0
5	Conv2D-5	256	(1, 1)	(1, 1)	[−1, 64, 56, 56]	4096
6	BatchNorm2d-6	64	(1, 1)	(1, 1)	[−1, 64, 56, 56]	128
7	ReLU-7	64	(3, 3)	(1, 1)	[−1, 64, 56, 56]	0
8	Conv2d-8	256	(1, 1)	(1, 1)	[−1, 64, 56, 56]	36,864
9	BatchNorm2d-9	64	(1, 1)	(1, 1)	[−1, 64, 56, 56]	128
10	ReLU-10	64	(1, 1)	(1, 1)	[−1, 64, 56, 56]	0
11	Bottleneck-512	256	(1, 1)	(1, 1)	[−1, 2048, 7, 7]	0
12	AdaptiveAvgPool2d-513	128	(1, 1)	(2, 1)	[−1, 2048, 1, 1]	0
13	Linear-514	128	(3, 3)	(1, 1)	[−1, 1000]	0
14	GlobalAveragePooling2D	–	–	–	(None, 2048)	0
15	Softmax Layer (Classification Layer)	–	–	–	(None, 1000)	0
16	Dense	–	–	–	(None, 1000)	2,049,000
			Total			**60,192,808**

$$PRE = \frac{(TP)}{(TP + FP)} \qquad (1.2)$$

$$REC = \frac{(TP)}{(TP + FN)} \tag{1.3}$$

$$F1 - S = \frac{(2 * PRE * REC)}{(PRE + REC)} \tag{1.4}$$

where TP (True Positives) is the number of correct positive predictions. TN (True Negatives) is the number of correct negative predictions. FP (False Positives) is the number of incorrect positive predictions. FN (False Negatives) is the number of incorrect negative predictions.

1.4   Experimental results and discussion

Figure 1.6 shows ResNet-152's accuracy for valence and arousal during training using the DEAP and DREAMER datasets. The model shows an accuracy of more

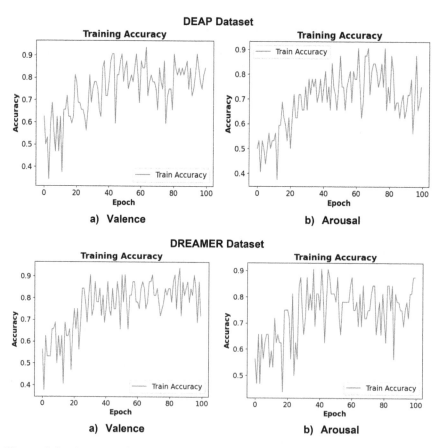

Figure 1.6   Accuracy during the ResNet-152 training process. (a) Valence emotion and (b) arousal emotion. The X-axis denotes the number of epochs, and the Y-axis denotes the accuracy.

than 90% at certain epochs in training. The accuracy starts increasing at the lower epoch value and stabilizes after a certain number of epochs. There is a similar performance for both datasets in terms of valence and arousal observed. The similar performance for both datasets underscores the model's consistency across diverse datasets. This consistent pattern may be due to the ResNet-152 architecture's capacity to detect complex features in EEG signals, contributing to its robust accuracy in emotion recognition tasks.

Loss functions are used to measure how well a proposed architecture responds to training data. During the training phase of our model, Figure 1.7 shows details of the loss function for the two datasets DEAP and DREAMER, as well as for the emotions valence and arousal. Variations seen in the loss function are a sign of how the model is learning, with modifications made to reduce errors and improve prediction accuracy. Since the model adjusts its parameters differently to maximize performance for each emotional dimension, the distinct patterns for valence and arousal highlight the complexity of emotion recognition. This technical perspective

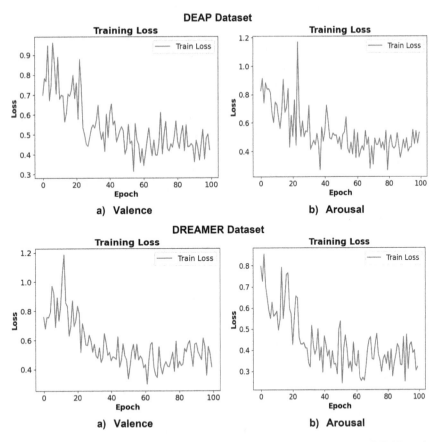

Figure 1.7   Loss function during ResNet-152 training process using DEAP and DREAMER dataset. (a) Valence and (b) arousal. The X-axis denotes the number of epochs, and the Y-axis denotes the loss function.

on the behavior of the loss function helps us better understand how the ResNet-152 model adapts to the complexity of EEG data for accurate emotion recognition across a variety of datasets and emotions.

Results for this study were obtained using subject-independent inference methods (SIR) and subject-correlated inference methods (SCR), while valence and arousal performance metrics were calculated using ResNet-152, ResNet152 +XGBoost, and ResNet152+ELM. Tables 1.1 and 1.2 show different models used to predict emotions for DEAP and DREAMER datasets using LOSO CV and subject-independent detection (SIR), respectively. In this study, three models were compared: ResNet-152, ResNet152+XGBoost, and ResNet152+ELM. Various metrics of valence and arousal were used to assess the performance of each model. In Tables 1.3 and 1.4, results from the same models are shown using a five-fold CV, also known as SCR method.

1.4.1   Experimental results and discussion using SIR method

Table 1.2 shows the accuracy of the ResNet-152 and ResNet152+XGBoost models is close to the same, while the ResNet152+ELM model shows a slightly higher accuracy of approximately 74%. In comparison with ResNet152+XGBoost, which has a precision of approximately 80%, and ResNet-152, which has a precision of approximately 79%, ResNet152+ELM has a precision of approximately 94% (approximately). Additionally, ResNet152+ELM's F1-S and REC are higher than the other two models. According to valence analysis of DREAMER data, ResNet152 +ELM has an accuracy of 87%, while ResNet152+XGBoost and ResNet-152 have accuracies of 85% and 83%, respectively. Other performance metrics like PRE, REC, and F1-S also show a higher value for the ResNet152+ELM model.

Table 1.3 illustrates that the accuracy of all three models, ResNet-152, ResNet152 +XGBoost, and ResNet152+ELM, for arousal using DEAP data is almost 80%. It appears that the DREAMER dataset has an accuracy of approximately 94% (approximately). In both datasets, ResNet-152 PRE is higher than ResNet152 +XGBoost and ResNet152+ELM PRE. In addition, REC and F1-S for ResNet-152 are higher in both datasets than for ResNet152+XGBoost and ResNet152+ELM.

1.4.2   Experimental results and discussion using SCR models

Table 1.4 shows the performance of emotion classification based on valence data. In this five-fold correlated model, the data are divided equally among five folds, and the average of the results is considered. DEAP dataset accuracy is 59% for ResNet152+ELM and 59% and 57% for ResNet-152 and ResNet152+XGBoost, respectively. Based on the DREAMER dataset, ResNet152+ELM model accuracy is 63%, ResNet152+XGBoost model accuracy is 67%, and ResNet152+ELM model accuracy is 63%. All three models have a PRE of 78%. ResNet-152 and ResNet152+ELM have PREs of 73%, while ResNet152+XGBoost has a PRE of 67%. In comparison to ResNet152+ELM, ResNet-152, and ResNet152+XGBoost, DREAMER has a REC of 73% and 68%, respectively.

For both datasets under consideration, the arousal scores for SCR-based models are shown in Table 1.5. ResNet-152 and ResNet152+ELM accuracy

Table 1.2   Performance (average±SD) performance (%) of emotion classification based on valence through SIR CV model of the DEAP and DREAMER dataset

Dataset	ResNet-152				ResNet152+XGBoost				ResNet152+ELM			
	ACC	PRE	REC	F1-S	ACC	PRE	REC	F1-S	ACC	PRE	REC	F1-S
DEAP	65.55±7.18	73.14±6.29	81.50±9.21	72.25±4.97	68.28±4.14	75.50±5.11	74.50±5.11	76.50±5.11	**69.92±4.09**	90.50±4.94	88.50±5.11	92.41±4.94
DREAMER	74.15±9.71	76.43±5.92	72.30±5.77	76.10±4.46	77.78±8.70	79.61±6.86	76.96±6.86	77.78±6.86	**80.19±6.66**	78.78±5.32	75.00±7.85	76.70±7.08

Note: The bold values represent the highest average accuracy scores.

Table 1.3   Performance (average ±SD) performance (%) of emotion classification based on arousal through SIR CV model of the DEAP and DREAMER dataset

Dataset	ResNet-152				ResNet152+XGBoost				ResNet152+ELM			
	ACC	PRE	REC	F1-S	ACC	PRE	REC	F1-S	ACC	PRE	REC	F1-S
DEAP	72.89±7.38	84.75±8.01	84.44±8.35	85.56±7.58	72.41±7.24	83.25±5.96	86.94±7.75	81.25±5.96	**73.87±5.59**	78.47±11.57	69.25±5.96	79.00±11.23
DREAMER	81.88±12.10	79.96±6.86	76.96±6.86	77.78±5.28	**82.85±10.71**	79.39±5.56	61.17±7.57	77.78±5.28	82.37±7.79	69.48±8.72	67.00±6.78	68.78±10.82

Note: The bold values represent the highest average accuracy scores.

Table 1.4 *Performance (average±SD) performance (%) of emotion classification based on valence through SCR CV model of the DEAP and DREAMER dataset*

Dataset	Fold	ResNet-152				ResNet152+XGBoost				ResNet152+ELM			
		ACC	PRE	REC	F1-S	ACC	PRE	REC	F1-S	ACC	PRE	REC	F1-S
DEAP	Fold-1	57.92	71.17	76.88	72.49	57.92	73.36	70.26	75.57	60.00	79.19	78.27	80.26
	Fold-2	58.33	79.86	66.42	80.32	61.67	84.20	81.18	85.61	59.17	82.33	83.33	81.45
	Fold-3	52.92	68.33	94.38	70.36	57.50	67.33	64.27	69.27	57.50	73.48	72.26	74.34
	Fold-4	55.00	90.15	81.11	92.17	57.50	91.12	89.81	94.48	59.29	78.87	79.19	77.27
	Fold-5	62.14	85.27	63.67	86.23	59.29	76.45	73.31	77.34	58.57	81.37	80.36	82.12
	Average	**57.26**	**78.60**	**76.00**	**80.06**	**58.77**	**78.20**	**75.40**	**80.08**	**58.90**	**78.60**	**78.40**	**78.80**
DREAMER	Fold-1	61.11	76.24	69.64	62.84	63.33	73.18	67.42	79.27	62.22	78.28	79.25	87.67
	Fold-2	68.89	63.02	58.34	79.52	70.00	59.56	72.26	62.22	64.44	63.36	61.43	72.34
	Fold-3	66.67	58.53	74.12	55.24	67.78	68.46	59.47	75.18	63.33	82.24	74.45	81.82
	Fold-4	56.67	71.36	61.03	67.61	58.89	75.18	64.56	68.48	62.22	68.26	70.16	76.26
	Fold-5	66.67	64.21	67.73	74.73	72.22	62.33	78.19	81.18	64.81	76.34	83.45	69.22
	Average	**64.00**	**66.67**	**66.17**	**67.99**	**66.44**	**67.40**	**68.00**	**73.00**	**63.41**	**73.42**	**73.53**	**77.00**

Note: The bold value represents the average performance score.

Table 1.5 Performance (average ±SD) performance (%) of emotion classification based on arousal through SCR CV model of the DEAP and DREAMER dataset

Dataset	Fold	ResNet-152				ResNet152+XGBoost				ResNet152+ELM			
		ACC	PRE	REC	F1-S	ACC	PRE	REC	F1-S	ACC	PRE	REC	F1-S
DEAP	Fold-1	58.75	90.16	89.23	91.45	57.92	95.23	94.21	96.14	55.83	61.14	57.14	63.36
	Fold-2	60.83	83.36	82.14	84.38	59.17	88.46	87.46	89.16	62.50	76.23	72.24	79.75
	Fold-3	64.58	67.33	68.45	66.33	67.50	62.13	63.32	61.28	67.08	70.45	67.13	71.43
	Fold-4	66.43	87.77	86.67	88.25	59.29	92.12	91.34	93.14	66.79	86.56	89.26	85.15
	Fold-5	60.00	91.13	90.17	92.14	59.29	96.31	95.21	97.23	56.93	64.23	62.91	66.18
	Average	62.12	83.6	83.00	84.20	60.63	86.60	86.00	87.20	62.23	71.4	69.4	72.82
DREAMER	Fold-1	70.00	86.17	81.27	86.17	72.22	88.67	87.77	86.32	71.11	87.16	88.18	89.49
	Fold-2	81.11	68.28	68.34	73.37	81.11	72.48	73.38	71.31	75.56	74.47	72.28	71.36
	Fold-3	75.56	75.45	76.32	79.45	75.56	79.16	80.45	78.17	74.44	81.28	80.19	82.26
	Fold-4	76.67	62.28	63.48	68.23	77.78	68.25	69.19	69.67	76.67	69.33	71.18	68.98
	Fold-5	72.22	81.14	85.25	82.28	74.07	85.36	84.16	83.17	72.22	85.46	86.19	84.34
	Average	75.11	74.43	74.6	77.6	76.15	78.4	78.60	77.40	73.48	79.43	79.20	78.18

percentages remain at the same level, while ResNet152+XGBoost accuracy percentages for the DEAP dataset are somewhat lower. There is almost no difference in accuracy between the models for the DREAMER dataset.

Overall, the findings show that on both the DEAP and DREAMER datasets, affect recognition tasks were successfully completed with high accuracy using ResNet-based models in combination with other classifiers. Diverse results across the various models demonstrate the impact of feature extraction and classifier selection.

1.5   Major limitations

Our study has several limitations. First, the dataset size and quantity, represented by DEAP and DREAMER, are relatively limited, limiting the generalizability of our results because diverse people experience emotions in different ways. Additionally, the models' performance in real-world situations, which can be influenced by noise, distractions, and individual differences that may affect EEG data quality and emotion expression, is a matter of concern given the emphasis on controlled experimental conditions. The difficulties in recognizing emotions come from their high subject variability, which may restrict the applicability of our suggested method. Finally, hybrid models' interpretability particularly that of ResNet-ELM is complicated, which presents issues in domains like healthcare, where transparency and clarity in understanding how models arrive at conclusions are essential for trust and reliable applications.

1.6   Future work

The acquisition of diverse datasets and the extension of model evaluation to real-world settings are critical steps toward the advancement of emotion recognition from EEG signals. The accuracy of the system is improved by multimodal approaches that incorporate speech, physiological signals, and facial expressions. Additionally, researching continuous, real-time emotion monitoring with EEG signals offers potential benefits in various domains and encourages further exploration of ResNet and machine learning techniques for advancing emotion identification, such as personalized user experiences and human–computer interactions.

1.7   Conclusion

Our research shows that ResNet combined with other machine learning models can greatly enhance the accuracy of emotion detection from EEG signals. Among the hybrid models we constructed, the ResNet-ELM model had the highest accuracy of 83.4%, outperforming the pure ResNet model. In this study, researchers demonstrated that ResNet-based models can be used to classify human emotions accurately using EEG data. We analyzed two distinct datasets, DEAP and DREAMER, which further support the validity and reliability of our findings. This study makes significant contributions to the field by introducing a novel approach that combines ResNet with various machine learning models to enhance emotion recognition accuracy. The inclusion of ResNet152-XGBoost, ResNet152-ELM, and ResNet-152 models in the comparison highlights how adaptable ResNet architectures are. It assures that our hybrid ResNet models perform better than pure ResNet models in recognizing emotions. Moreover, the way we categorize emotional states into valence and arousal adds a deeper level of understanding. Thus, this work stands out for its high accuracy and methodical comparison of hybrid models, which reveal differences in their respective performances.

References

[1]   D. H. Hockenbury and S. E. Hockenbury, *Discovering Psychology*, Macmillan, 2010.

[2]   M. Murugappan, M. Rizon, R. Nagarajan, and S. Yaacob, "FCM clustering of human emotions using wavelet based features from EEG," *Int J Biomed Soft Comput Human Sci: The Official J Biomed Fuzzy Syst Assoc*, vol. 14, no. 2, pp. 35–40, 2009.

[3]   M. Murugappan, "Frequency band localization on multiple physiological signals for human emotion classification using DWT," in A. Konar and A. Chakraborty (eds), *Emotion Recognition: A Pattern Analysis Approach*, pp. 295–313. Hoboken, NJ: John Wiley & Sons, Inc., 2015.

[4]   P. Ekman, "An argument for basic emotions," *Cogn Emot*, vol. 6, no. 3–4, pp. 169–200, 1992.

[5]   R. W. Picard, E. Vyzas, and J. Healey, "Toward machine emotional intelligence: Analysis of affective physiological state," *IEEE Trans Pattern Anal Mach Intell*, vol. 23, no. 10, pp. 1175–1191, 2001.

[6]   I. A. Essa and A. P. Pentland, "Coding, analysis, interpretation, and recognition of facial expressions," *IEEE Trans Pattern Anal Mach Intell*, vol. 19, no. 7, pp. 757–763, 1997.

[7]   M. Song, C. Chen, J. Bu, and M. You, "Speech emotion recognition and intensity estimation," in *Computational Science and Its Applications–ICCSA 2004: International Conference, Assisi, Italy, May 14–17, 2004, Proceedings, Part IV 4*, Springer, 2004, pp. 406–413.

[8]   H. J. Yoon and S. Y. Chung, "EEG-based emotion estimation using Bayesian weighted-log-posterior function and perceptron convergence algorithm," *Comput Biol Med*, vol. 43, no. 12, pp. 2230–2237, 2013.

[9]   M. Murugappan, M. Rizon, R. Nagarajan, A. S. AlMejrad, and S. Yaacob, "Comparison of human emotion classification through different set of EEG channels," in *15th International Symposium on Artificial Life and Robotics, AROB'10*, 2010, pp. 216–220.

[10] M. Murugappan and S. Yaacob, "Asymmetric ratio and FCM based salient channel selection for human emotion detection using EEG," *WSEAS Trans Signal Process*, 2008.

[11] C. Yu and M. Wang, "Survey of emotion recognition methods using EEG information," *Cogn Robot*, vol. 2, pp. 132–146, 2022.

[12] S. Koelstra, A. Yazdani, M. Soleymani, *et al.*, "Single trial classification of EEG and peripheral physiological signals for recognition of emotions induced by music videos," in *Proceedings of Brain Informatics: International Conference, BI 2010,* Toronto, ON, Canada, August 28–30, 2010. Springer, 2010, pp. 89–100.

[13] P. J. Lang, M. M. Bradley, and B. N. Cuthbert, "Emotion, attention, and the startle reflex," *Psychol Rev*, vol. 97, no. 3, p. 377, 1990.

[14] D. M. Tucker, M. Liotti, G. F. Potts, G. S. Russell, and M. I. Posner, "Spatiotemporal analysis of brain electrical fields," *Hum Brain Mapp*, vol. 1, no. 2, pp. 134–152, 1994.

[15] X.-W. Wang, D. Nie, and B.-L. Lu, "Emotional state classification from EEG data using machine learning approach," *Neurocomputing*, vol. 129, pp. 94–106, 2014.

[16] S. Koelstra, C. Muhl, M. Soleymani, *et al.*, "DEAP: A database for emotion analysis; using physiological signals," *IEEE Trans Affect Comput*, vol. 3, no. 1, pp. 18–31, 2012.

[17] Y. LeCun, Y. Bengio, and G. Hinton, "Deep learning," *Nature*, vol. 521, no. 7553, pp. 436–444, 2015.

[18] R. Sharma, R. B. Pachori, and P. Sircar, "Automated emotion recognition based on higher order statistics and deep learning algorithm," *Biomed Signal Process Control*, vol. 58, p. 101867, 2020.

[19] L. Shu, J. Xie, M. Yang, *et al.*, "A review of emotion recognition using physiological signals," *Sensors*, vol. 18, no. 7, p. 2074, 2018.

[20] A. Craik, Y. He, and J. L. Contreras-Vidal, "Deep learning for electroencephalogram (EEG) classification tasks: A review," *J Neural Eng*, vol. 16, no. 3, p. 031001, 2019.

[21] R. Yuvaraj, A. Baranwal, A. A. Prince, M. Murugappan, and J. S. Mohammed, "Emotion recognition from spatio-temporal representation of EEG signals via 3D-CNN with ensemble learning techniques," *Brain Sci*, vol. 13, no. 4, p. 685, 2023.

[22] M. Maithri, U. Raghavendra, A. Gudigar, *et al.*, "Automated emotion recognition: Current trends and future perspectives," *Comput Methods Programs Biomed*, vol. 215, p. 106646, 2022.

[23] A. Hassouneh, A. M. Mutawa, and M. Murugappan, "Development of a real-time emotion recognition system using facial expressions and EEG based on machine learning and deep neural network methods," *Inform Med Unlocked*, vol. 20, p. 100372, 2020.

[24] T. Song, W. Zheng, P. Song, and Z. Cui, "EEG emotion recognition using dynamical graph convolutional neural networks," *IEEE Trans Affect Comput*, vol. 11, no. 3, pp. 532–541, 2018.

[25] D. Nie, X.-W. Wang, L.-C. Shi, and B.-L. Lu, "EEG-based emotion recognition during watching movies," in *2011 5th International IEEE/EMBS Conference on Neural Engineering*, IEEE, 2011, pp. 667–670.

[26] Y.-J. Liu, M. Yu, G. Zhao, J. Song, Y. Ge, and Y. Shi, "Real-time movie-induced discrete emotion recognition from EEG signals," *IEEE Trans Affect Comput*, vol. 9, no. 4, pp. 550–562, 2017.

[27] H. Lee and S. Lee, "Arousal-valence recognition using CNN with STFT feature-combined image," *Electron Lett*, vol. 54, no. 3, pp. 134–136, 2018.

[28] M. Murugappan, "Human emotion classification using wavelet transform and KNN," in *2011 International Conference on Pattern Analysis and Intelligence Robotics*, IEEE, 2011, pp. 148–153.

[29] K. Veropoulos, C. Campbell, and N. Cristianini, "Controlling the sensitivity of support vector machines," in *Proceedings of the International Joint Conference on AI*, Stockholm, 1999, p. 60.

[30] K. Guo, R. Chai, H. Candra, *et al.*, "A hybrid fuzzy cognitive map/support vector machine approach for EEG-based emotion classification using compressed sensing," *Int J Fuzzy Syst*, vol. 21, pp. 263–273, 2019.

[31] A. H. Krishna, A. B. Sri, K. Y. V. S. Priyanka, S. Taran, and V. Bajaj, "Emotion classification using EEG signals based on tunable-Q wavelet transform," *IET Sci Measure Technol*, vol. 13, no. 3, pp. 375–380, 2019.

[32] W.-L. Zheng, J.-Y. Zhu, Y. Peng, and B.-L. Lu, "EEG-based emotion classification using deep belief networks," in *2014 IEEE International Conference on Multimedia and Expo (ICME)*, IEEE, 2014, pp. 1–6.

[33] S. K. Hadjidimitriou and L. J. Hadjileontiadis, "Toward an EEG-based recognition of music liking using time–frequency analysis," *IEEE Trans Biomed Eng*, vol. 59, no. 12, pp. 3498–3510, 2012.

[34] R. Yuvaraj, P. Thagavel, J. Thomas, J. Fogarty, and F. Ali, "Comprehensive analysis of feature extraction methods for emotion recognition from multi-channel EEG recordings," *Sensors*, vol. 23, no. 2, p. 915, 2023.

[35] S. Li, J. Jiao, Y. Han, and T. Weissman, "Demystifying ResNet," *arXiv preprint arXiv:1611.01186*, 2016.

[36] R. U. Khan, X. Zhang, and R. Kumar, "Analysis of ResNet and GoogleNet models for malware detection," *J Comput Virol Hack Tech*, vol. 15, pp. 29–37, 2019.

[37] K. He, X. Zhang, S. Ren, and J. Sun, "Deep residual learning for image recognition," in *Proceedings of the IEEE Conference on Computer Vision and Pattern Recognition*, 2016, pp. 770–778.

[38] T. Chen and C. Guestrin, "Xgboost: A scalable tree boosting system," in *Proceedings of the 22nd ACM SIGKDD International Conference on Knowledge Discovery and Data Mining*, 2016, pp. 785–794.

[39] T. Tian, L. Wang, M. Luo, Y. Sun, and X. Liu, "ResNet-50 based technique for EEG image characterization due to varying environmental stimuli," *Comput Methods Programs Biomed*, vol. 225, p. 107092, 2022.

[40]   A. Topic and M. Russo, "Emotion recognition based on EEG feature maps through deep learning network," *Int J Eng Sci Technol*, vol. 24, no. 6, pp. 1442–1454, 2021.

[41]   S. Katsigiannis and N. Ramzan, "DREAMER: A database for emotion recognition through EEG and ECG signals from wireless low-cost off-the-shelf devices," *IEEE J Biomed Health Inform*, vol. 22, no. 1, pp. 98–107, 2017.

[42]   C. Cortes and V. Vapnik, "Support-vector networks," *Mach Learn*, vol. 20, pp. 273–297, 1995.

[43]   G.-B. Huang, Q.-Y. Zhu, and C.-K. Siew, "Extreme learning machine: theory and applications," *Neurocomputing*, vol. 70, no. 1–3, pp. 489–501, 2006.

[44]   S. Hochreiter and J. Schmidhuber, "Long short-term memory," *Neural Comput*, vol. 9, no. 8, pp. 1735–1780, 1997.

Chapter 2

Detection of facial emotion using thermal imaging based on deep learning techniques

B. Sathyamoorthy[1], U. Snekhalatha[1] and T. Rajalakshmi[2]

Facial emotion recognition is one of the affective computing tasks in computer vision. Human emotion recognition based on facial expression is a challenging task in the field of involvement of intelligent systems in human–computer interaction. Nevertheless, some limitations have been found in the present scenario of facial emotion detection, such as the network model's inability to generalize, and the recognition system's inability to be robust. In this article, we propose a method for detecting facial emotions using deep neural networks such as MobileNet-V2 and DenseNet-121 using digital and thermal imaging. The study's goals are to (1) determine the various emotions, such as anger, happiness, neutrality, and sadness, from digital and thermal images of different human subjects' faces, (2) compare the data, and (3) validate the performance analysis. The accuracy value was obtained from digital and thermal images. Fifty normal subjects of both genders aged 20±3 years were included in the proposed study. Digital and thermal images for four emotions with a total of 200 digital and 200 thermal images were obtained. The modified DenseNet-121 generated the highest test accuracy of 94.9% and 95.1% using digital and thermal images, respectively, in comparison to MobileNet-V2 which produced only 82.1% and 91.4% accuracy for the classification of facial emotions. It has been shown that the proposed DenseNet-121 model has better discrimination power in recognizing anger, happiness, neutrality, and sad facial emotions from digital and thermal images. The proposed study showed the relevant demonstration for better performance of thermal imaging compared to the digital imaging method in facial expression detection.

Keywords: Deep residual network; Digital imaging; Thermal imaging; Emotion recognition; Facial expression recognition; MobileNet-V2

[1]Department of Biomedical Engineering, College of Engineering, and Technology, SRM Institute of Science, and Technology, India
[2]Department of Electronics, and Communication Engineering, College of Engineering, and Technology, SRM Institute of Science, and Technology, India

2.1   Introduction

Automated emotion detection in facial regions finds an extensive spectrum of applications in computer vision and robotics, especially in the fields of social signal processing and affective computing. The main objective of automated facial expression recognition (FER) is to recognize and categorize individual facial expressions into related emotions [1]. The extraction of facial features is a highly sought-after implementation in the field of surveillance (video or pictures), biometrics, and human–computer interface.

The universal emotions recognized by facial expression are neutral, angry, disgust, fear, happy, sad, and surprise. Because the facial expressions of every individual are unique, recognizing facial features is a challenging task. Physical characteristics such as sex, genes, and age all have an impact on the extraction of facial features. Hence there is a real need to develop an automated emotion and facial expression identification system to recognize and classify emotions and facial expressions more accurately considering all the vital parameters. The developed FER system must be able to recognize facial expressions in a variety of situations, such as changes in light source, varying illuminations, spectacle usage, the presence of facial hair, and so on. These are the critical issues that need to be addressed in the current scenario.

Usually for FER, digital cameras are used for the image acquisition process. However, due to poor illumination of light, reflection, and shadowing effects, the categorization of various facial emotions is a challenging task. Hence thermal imaging modality is a better choice for studying the skin surface temperature distribution, and thermal characteristics in the facial muscles, and for the classification of facial expressions. The thermal imaging method is used in a variety of applications, which include fever detection in screening COVID-19 subjects [2], human–robot interaction [3], and orofacial detection [4] in facial thermal images.

Hence the aim and objectives of the proposed study are: (i) To determine the various emotions namely happy, angry, neutral, and sad from the acquired facial images of subjects using digital and thermal cameras. (ii) To classify the different emotions using deep convolutional neural network (CNN) models like MobileNet-V2 and DenseNet-121, and to compare the performance of digital and thermal imaging techniques based on the accuracy.

2.2   Related works

Recognition of facial emotions is considered to be one of the most challenging tasks. Facial emotions, in general, rely on the contemplation of the person's mind which shows various forms of emotions, such as happiness, neutral, sad, fear, surprise, anger, and disgust. Hence categorization of facial emotion sounds to be a challenging area in cognitive computation. Khattak *et al.* [5] developed a CNN to classify emotions based on the facial expressions to determine the age and gender of the subject under study. They obtained an accuracy of 95.65%, 98.5%, and 99.14% for emotion, age, and gender recognition, respectively.

Mehendal *et al.* proposed a facial emotion recognition technique using a CNN model [6]. The developed model has two parts: the first part will perform background elimination, and the second part focuses on the determination of facial feature vectors through which different facial expressions are estimated. The study resulted in an accuracy value of 96%. Wahyono *et al.* [7] developed a hybrid computational intelligence algorithm for facial emotion detection based on facial expressions. The author implemented a Fuzzy C-Means algorithm to group colour that enables face detection. Histograms were obtained upon colour grouping. Numeric values of each facial emotion were determined using the obtained histogram. The data set was formulated by the numerical value obtained from the histogram. They attained an accuracy of 77.5% using this ubiquitous device.

Mollahosseini *et al.* [8] performed emotion recognition based on deep learning techniques using seven publicly available databases. The author used single-component CNN architecture, designed a custom model, and compared it with the Alex net model. Hence the author was able to prove that the proposed architecture was able to recognize facial emotions and attained an accuracy of 89.3%.

Mukhopadhyay *et al.* [9] proposed a method to predict facial emotions using textural images based on a deep learning model. The textural features of the human faces were extracted using the local binary pattern (LBP) method. The authors used the Extended Cohn-Kanade (CK+) dataset in their study to train the CNN. The accuracy level of the developed technique was compared with the existing conventional CNN. Kola and Samayamantula [10] developed an approach for feature extraction to recognize facial emotions. In this study, authors calculated the LBP taking into account four neighbours, and diagonal neighbours, an adaptive window, and averaging in radial directions were introduced for feature description. Facial expressions were classified using a support vector machine. Furthermore, the performance of the proposed algorithm was assessed using a recognition rate and confusion matrix.

Bhattacharyya *et al.* [11] collected a total of 1782 samples of various human facial emotions from the IR database. They developed the IRFacExNet model to categorize facial expressions using thermal imaging techniques. They trained their net with a batch size of 16 samples and obtained only 82.83% accuracy with the naïve model. However, the authors obtained a higher accuracy of 88.43% with the help of an ensemble system of snapshot models for the classification of human facial emotions.

Elbarawy *et al.* [12] implemented FER based on deep learning techniques using thermal images. They compared the performance of the neural network (NN), deep autoencoder neural networks (AENN), and CNN for the categorization of facial expressions using digital and thermal imaging. They used the IRIS dataset that contains 30 thermal and digital images of 28 male and 2 female subjects. They conducted their study on three different facial expressions, namely, surprise, happy, and angry. The AENN produced 90% accuracy followed by CNN with 93.3% accuracy.

2.3   Materials and methodology

2.3.1   Participants

A total set of 50 healthy subjects of each gender was considered for this study to analyze the facial expressions between the digital and thermal images. Facial expressions for various emotions such as neutral, happy, sad, and angry were acquired simultaneously using digital and thermal cameras. Four emotions for 50 subjects on an average of 200 images were analyzed independently for the imaging modalities. The study was carried out at the Biomedical Engineering Department, Kattankulathur campus, Chennai, Tamil Nadu, India from October 2021. The subjects under the age group of 20–30 years are the inclusion criteria for this study. This study excludes participants above the age group of 30 years, as well as those who are suffering from illnesses such as a cold, cough, fever, and respiratory infections. Approval for human ethical clearance is provided by the ethical committee of SRM Medical College Hospital and Research Centre, Kattankulathur, Tamil Nadu, India with approval number 2992/IEC/2021.

2.3.2   Image acquisition

The participants were asked to sit down on a chair comfortably and made to rest for a while before commencing the experimental procedure. The image acquisition room is provided with temperature controlled with proper lighting for acquiring digital images and a dark ambience room for acquiring thermal images. The subjects were asked to remove external accessories, such as spectacles and masks, and were instructed to sit in an upright position. They were asked to keep their heads in a fixed position and avoid any sudden movement. The thermal imaging study was carried out using a FLIR camera (A305SC) having an IR resolution of 320 × 240.

2.3.3   Dataset

The acquired data of 200 digital and 200 thermal images representing 50 images each for angry, happy, neutral, and sad emotions for both sets of digital and thermal images. Data augmentation techniques such as rotation, translation, flipping, and scaling were incorporated in the study. Through the data augmentation technique, 700 digital and 700 thermal images were obtained. Further incorporating patch extraction techniques, the dataset was increased to 7000 digital and 7000 thermal images. For both the imaging modalities, training the network was incorporated using 4900 images (70% from the original dataset), 2100 images (30%) were used for validation, and 1050 images were used as testing data.

Data from digital and thermal images of faces labelled with their respective expressions are used to train the network. The captured images are subjected to pre-processing that could enable to identification of the required region of interest. A data augmentation technique is adopted to boost the amount of data for classification purposes. Furthermore, 70% of the dataset is subjected to training and 30% for validation. Deep learning models such as DenseNet and MobileNet-V2 are

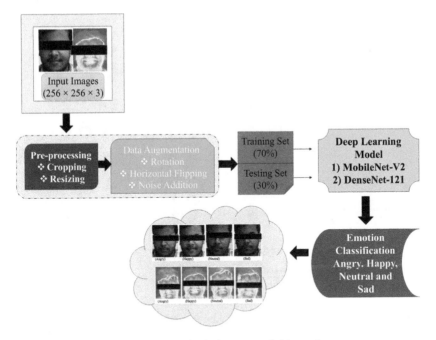

Figure 2.1 Block diagram of this study

proposed for the classification of different facial emotions using both imaging modalities independently. Figure 2.1 represents the overall block diagram for the proposed study facial emotion detection using thermal imaging.

2.3.4 MobileNet-V2

MobileNet is one of the most successful CNN architectures and requires less memory while still providing superior predictions [13]. It is based on an inverted residual structure where the residual connections are between the bottleneck layers. To train an algorithm, Google Engineers used data from their servers. This network is ideal for devices with lower processing power, such as mobile devices.

MobileNet-V1 consists of two layers such as depth-wise convolution and point-wise convolution layer. The depth-wise convolution layer is used for filtering using a convolutional filter, and the point-wise layer is used for constructing new features based on the linear combinations of input channels. MobileNet-V2 is an improved version of MobileNet-V1. It is comprised of two residual blocks in which one residual block has a stride length of 1, and the other residual block has a stride length of 2 [14–16]. Each of the residual blocks is comprised of three layers a 1×1 convolution layer with RELU activation function, a depth-wise convolution layer, and another 1×1 convolutional layer.

In order to classify all the emotions found in faces, we used a classifier that categorizes them into four groups: angry, happy, neutral, and sad. Because of its high accuracy and low average inference time, the MobileNet-V2 was chosen.

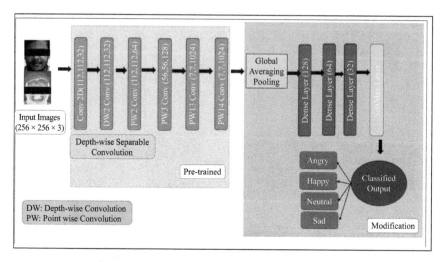

Figure 2.2   Digital and thermal images of a subject's facial expressions with a modified MobileNet-V2 architecture

A detector connected to MobileNet's backbone network achieved the highest detection speed. MobileNet-V2 consists of bottleneck residual networks which contain 19 basic blocks. 1×1 convolution layer with an average pooling layer is succeeding the basic blocks. Convolutions are performed depth-wise instead of combining all three and flattening them, so each colour channel gets its convolution. It allows filtering of the input channels. By using depth-wise separable convolution, the complication and model size of the network can be decreased. After that dense blocks are connected which uses the layers by changing or reducing the dimension of a vector by neurons. A softmax activation function is used at the final stage of classification to perceive the different emotions from the digital and thermal images. Figure 2.2 demonstrates the modified MobileNet-V2 architecture for the categorization of various facial emotions.

2.3.5   DenseNet-121 architecture

Deep neural networks based on DenseNet-121 [17] overcome the vanishing gradient problem for extremely deep convolutional networks. Concatenation of multi-layer features is performed by DenseNet-121, a CNN-based architecture. In dense networks, these blocks consist of 6, 12, 24, and 16 layers each, because all layers can communicate with each other, thereby enabling the reuse of feature sets. In each subsequent layer, feature maps are shifted from the previous layer, and inputs from the preceding layer are obtained. In addition, these networks are quite narrow by nature, with an average of 12 filters used for the extraction of features. Because the narrow architecture makes the layers directly accessible to the gradient of the error signal, the training performance is improved, in turn resulting in good parameter efficiency. This network is made up of two essential computing configurations: (a) dense block and (b) transitive layer. After learning these high-level features, the fully connected layers are used. Feature reuse reinforces feature propagation,

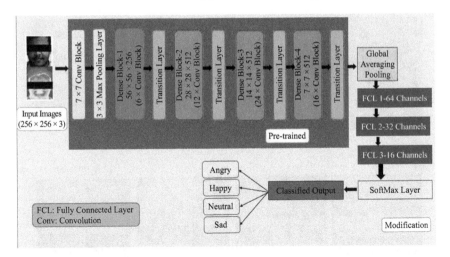

Figure 2.3   Digital and thermal images of a person's facial expressions with a modified DenseNet-121 architecture

enhances feature propagation, reduces the vanishing-gradient problem, and significantly reduces parameter number are the main advantages of DenseNet-121.

The fully connected layer of DenseNet-121 has been modified to have two 1024 neuron feedforward NNs with four neurons each to apply it to the classification of facial expressions in digital, and thermal images. DenseNet-121 has 120 convolutions and 4 AvgPool. Each layer's feature map is concatenated to the input of every subsequent layer within a dense block and this connection is done using concatenation. This allows the next layers to directly access the features from previous layers, and also it allows the reuse of features within the network. The building block of DenseNet-121 is the dense blocks. As shown in Figure 2.3, each dense block has multiple convolution layers. A dense block is followed by a transition layer and even its output can proceed to the next dense block. The updated global average pooling layer accepts all the feature maps of the network and is trained to recognize face expressions using digital and thermal data. As part of the classification layer, we used softmax activation to detect digital and thermal facial emotion images. Figure 2.3 depicts a DenseNet-121 architecture that has been developed to detect facial expressions from digital and thermal images.

2.3.6   Parameters

We have used stochastic gradient descent (optimization algorithm) R, random sampling in log space to select the parameters (patch extraction, learning rate, decay, and momentum) that provided the best validation accuracy. Figure 2.4 shows the layers that we have changed for the deep learning network for both DenseNet-121 and MobileNet-V2 architecture. The hyperparameters used in the networks are learning rate as 0.01, learning rate decay as 1e-6, and momentum as 0.9. The number of patches used is 10, the batch size adopted is 10, and the number of epochs used is 50.

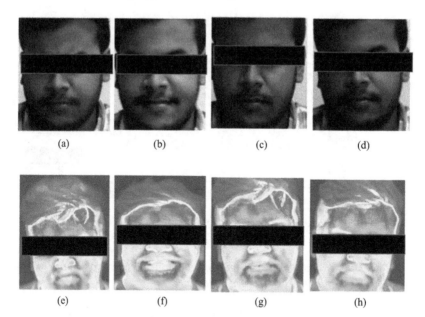

| (a) | (b) | (c) | (d) |

| (e) | (f) | (g) | (h) |

*Figure 2.4   Four facial expressions were considered for digital images, namely
            (a) angry, (b) happy, (c) neutral, (d) sad, and thermal images, namely
            (e) angry, (f) happy, (g) neutral, and (h) sad*

Table 2.1   Performance evaluation metrics of formulas

Evaluation metrics	Formula
Accuracy	$\frac{TP + TN}{TP + TN + FP + FN}$
Precision	$\frac{TP}{TP + FP}$
Recall	$\frac{TP}{TP + FN}$
F1-score	$2 \times \frac{Precision \times recall}{Precision + recall}$
Sensitivity	$\frac{TP}{TP + FN}$
Specificity	$\frac{TN}{TN + FP}$
Positive predictive value	$\frac{TP}{TP + FP}$
Negative predictive value	$\frac{TN}{TN + FN}$

2.3.7   Evaluation using a performance matrix

Performance measures such as accuracy, precision, recall, F1-score, sensitivity, specificity, positive predictive value, and negative predictive value are obtained for the
MobileNet-V2 and DenseNet-121 networks for the recognition of face emotions using
digital and thermal images. The formula for calculating the performance measures is
tabulated in Table 2.1. The area under the curve (AUC) value of the receiver operating
characteristic (ROC) curve is obtained for all the deep learning networks.

2.4 Results

The proposed study aims to perform automatic classification of facial emotions of thermal images using deep learning NN models such as MobileNet-V2 and DenseNet-121. The study involved four different emotions such as neutral, angry, happy, and sad. The individuals were subjected to videos of different emotions, and their facial expressions on the video were captured using a thermal camera. A similar procedure was implemented using a digital camera.

Figure 2.4 depicts various emotions captured using digital and thermal cameras, where (a) represents angry facial emotion for the digital images; (b) shows facial emotion during happy ambience for the digital images; (c) depicts the neutral facial emotion for the digital images; and (d) indicates sad facial emotion for the digital images. Similarly, in Figure 2.4, (e) indicates the angry emotion, (f) shows the happy emotion, (g) represents the neutral, and (h) represents the sad state of facial expressions for different emotions using thermal imaging.

A total of 4900 and 2100 images are used for training and validation for the proposed deep learning models, such as MobileNet-V2 and DenseNet-121 CNN models. For thermal images, the classifier achieved an accuracy of 91.4% and 95.1% using the MobileNet-V2 and DenseNet-121, respectively. Similarly, for digital images, the classifier achieved an accuracy of 81.2% and 94.9% using the MobileNet-V2 and DenseNet-121, respectively. Table 2.2 shows the confusion

Table 2.2 Confusion matrix for the MobileNet-V2 and DenseNet-121 CNN models using digital and thermal images

DNN	Images	Emotions	TP	FP	FN	TN	Sen	Spe	PPV	NPV	Acc
MobileNet-V2	Digital	Angry	760	12	109	175	87.45	93.58	98.44	61.61	88.54
		Happy	697	97	33	229	95.47	70.24	87.78	87.40	87.68
		Neutral	762	40	10	244	98.70	85.91	95.01	96.06	92.26
		Sad	761	39	36	220	95.48	84.94	95.12	85.93	92.89
	Thermal	Angry	784	14	23	235	97.14	94.37	98.24	91.08	96.49
		Happy	801	5	11	239	98.64	97.95	99.37	95.6	98.39
		Neutral	711	67	10	268	98.61	80	91.38	96.40	95.70
		Sad	782	4	46	224	94.4	98.24	99.49	82.96	95.26
DenseNet-121	Digital	Angry	781	13	14	248	98.23	95.01	98.36	94.65	97.44
		Happy	790	5	12	249	98.50	98.03	99.37	95.40	98.39
		Neutral	770	22	14	250	98.21	91.91	97.22	94.69	96.59
		Sad	774	13	13	256	98.34	95.16	98.34	95.16	97.53
	Thermal	Angry	774	10	9	263	98.85	96.33	98.72	96.69	98.20
		Happy	786	12	12	246	98.49	95.34	98.49	95.34	97.72
		Neutral	775	18	12	251	98.47	93.30	97.73	95.43	97.15
		Sad	781	12	19	244	97.62	95.31	98.48	92.77	97.06

Sen = Sensitivity, Spe = Specificity, Acc = Accuracy, PPV = Positive predictive value, NPV = Negative predictive value.

Table 2.3   Performance comparison between the digital and thermal images of four emotions in MobileNet-V2 and DenseNet-121 classifiers

Model	Images	Emotions	Precision	Recall	F1-score	Overall accuracy	AUC
MobileNet-V2	Digital	Angry	0.94	0.62	0.74	82.1	0.96
		Happy	0.70	0.87	0.78		
		Neutral	0.86	0.96	0.91		
		Sad	0.85	0.86	0.85		
		Weighted average	0.84	0.82	0.82		
	Thermal	Angry	0.94	0.91	0.93	91.4	0.99
		Happy	0.98	0.96	0.97		
		Neutral	0.80	0.96	0.87		
		Sad	0.98	0.83	0.90		
		Weighted average	0.92	0.91	0.92		
DenseNet-121	Digital	Angry	0.95	0.95	0.95	94.9	0.99
		Happy	0.98	0.95	0.97		
		Neutral	0.92	0.95	0.93		
		Sad	0.95	0.95	0.95		
		Weighted average	0.95	0.95	0.95		
	Thermal	Angry	0.96	0.97	0.97	95.1	0.99
		Happy	0.95	0.95	0.95		
		Neutral	0.93	0.95	0.94		
		Sad	0.95	0.93	0.94		
		Weighted average	0.95	0.95	0.95		

matrix for MobileNet-V2 and DenseNet-121 CNN models using digital and thermal images.

The performance measures of MobileNet-V2 and DenseNet-121 for identifying facial emotion using digital and thermal images are shown in Tables 2.2 and 2.3, respectively. MobileNet-V2 and DenseNet-121 CNN models were tested to determine their performance in terms of precision, recall, F1-score, overall accuracy, and AUC. The final measurements were derived using the mean of each of the parameters for the normal human subjects of digital and thermal face emotions. The MobileNet-V2 model of thermal image emotions demonstrated an overall accuracy of 91.4% with a weighted average precision of 0.92, recall (0.91), and F1-score (0.92). In terms of weighted average accuracy, the DenseNet-121 achieved a precision value of 0.95, a recall value of 0.95, and an F1-score of 0.95 using thermal images. Compared to the MobileNet-V2, DenseNet-121 images were 94.9% more accurate in digital and 95.1% more accurate in thermal imaging.

Thermal image emotions yielded an AUC of 0.99, but digital images yielded an AUC of 0.96 for the MobileNet-V2 model. The DenseNet-121's AUC was the same for digital and thermal images 0.99, respectively. Thermal images of four emotions performed much better than digital images on both networks. DenseNet-121 networks of thermal image emotions including angry, happy, neutral, and sad had an AUC of 0.99, whereas MobileNet-V2 had an AUC of 0.99 for digital images of four emotions. Compared to the MobileNet-V2, the DenseNet-121 model provides better overall assessment metrics. Thus, the DenseNet-121 network outperformed the MobileNet-V2 model in terms of accuracy. Figure 2.5(a) and (b) shows the 50 epochs of MobileNet-V2 training and validation performance curves for digital and thermal image emotions. Figure 2.5(c) and (d) shows the training and validation curves for DenseNet-121 for 50 epochs in the digital and thermal image emotions, respectively.

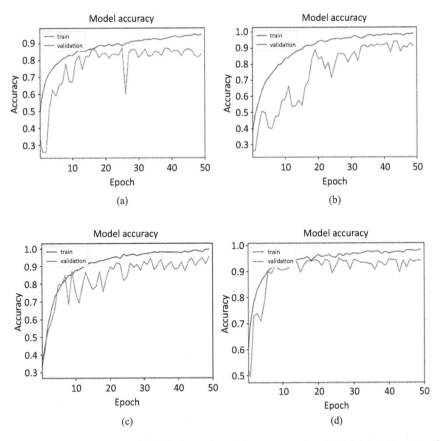

Figure 2.5   Training and validation of model accuracy for CNN MobileNet-V2 of
(a) digital images and (b) thermal images, and DenseNet-121 of
(c) digital images and (d) thermal images

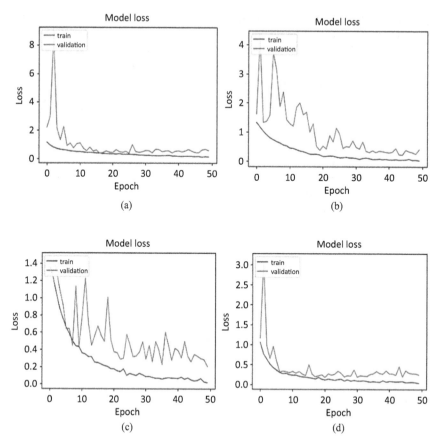

*Figure 2.6   Training and validation model loss for MobileNet-V2 of (a) digital
images and (b) thermal images, and DenseNet-121 of (c) digital
images and (d) thermal images*

MobileNet-V2 networks of digital and thermal image emotions of training
and validation loss performance curves are shown in Figure 2.6(a) and (b). The
training and validation model loss performance curves of DenseNet-121 digital
and thermal image emotions are shown in Figure 2.6(c) and (d). As shown in
Figure 2.7(a) and (b), the ROC curve for the MobileNet-V2 model for digital
and thermal image emotions reveals the ROC values of thermal image emotions
of 0.99 is high, while the ROC values of 0.96 for a similar range of emotional
states for digital images. The ROC curve for the DenseNet-121 model for digital
and thermal images is also shown in Figure 2.7(a,b), which demonstrates that
the area under the ROC obtained as 0.99 for various emotions using thermal
image is identical to that of the area under the ROC (0.99) obtained for digital
images.

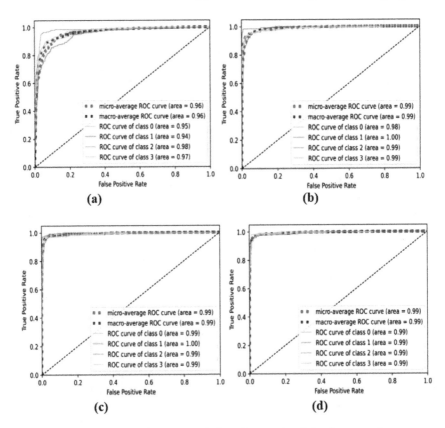

Figure 2.7   *ROC curve of supervised classifiers trained for face emotion recognition from MobileNet-V2 of (a) digital images and (b) thermal images, and DenseNet-121 of (c) digital images and (d) thermal images*

2.5 Discussion

Our study examined 50 normal human subjects' digital and thermal images to study the characteristics of specific types of emotions, such as anger, happiness, neutrality, and sadness. Using deep learning models based on CNN, this paper explores the recognition of facial expressions or emotions based on digital and thermal images. Digital and thermal imaging techniques were used to compare MobileNet-V2 and DenseNet-121 models in detecting and classifying angry, happy, neutral, and sad emotions. Using digital and thermal images, modified DenseNet-121 generated the highest test accuracy of 94.9% and 95.1% in comparison to MobileNet-V2 which produced only 82.1% and 91.4% accuracy for the classification of facial emotions, respectively. It has been shown that the proposed DenseNet-121 model has better discrimination power in recognizing angry, happy, neutral, and sad facial emotions from digital and thermal images.

A study conducted by Bodavarapu *et al.* [18] has proposed a new model that has a high performance on low-resolution and low-reliable images. The authors formed a dataset with low-resolution facial expression images (LRFE) consisting of 35,000 images as well as used the FER2013 dataset with 35,887 images that are publicly available online. The mixed dataset is composed of R, random images from FER2013 and LRFE datasets from each emotion category. Approximately 35,000 images in this mixed dataset correspond to seven different expressions (happiness, sadness, neutrality, fear, disgust, anger, and surprise). In addition, the dataset is split 80:20 into training and testing, and each deep neural network is run with 32 batches and 100 epochs. For DenseNet-121, FER2013 and LRFE achieve an accuracy of 60% and 68%, respectively. The hybrid filtering method combined with DenseNet-121 resulted in 95% accuracy. The hybrid filtering method also worked well for ResNet-50, MobileNet, and Xception models. Using this novel model, CNN obtained an accuracy of 65% on the FER2013 dataset, whereas other current models such as ResNet-50, MobileNet, DenseNet-121, and Xception achieved an accuracy of 60%, 57%, 60%, and 52%, respectively, on the same dataset. Because of combining the hybrid filtering approach with the suggested model, the precision yielded 85%.

Shirisha *et al.* illustrated a CNN using 48 × 48 pixel grayscale images in the FER2013 dataset as input [19]. To classify a person's emotional state, physiological signals are compared with this dataset. A digital image is pre-processed through methods such as face detection, RGB2gray conversion, and resizing. Face detection is done by using the Haar cascade classifier, and then the image frame is converted into grayscale, resized and reshaped to 48 × 48 pixels. A process of data pre-processing is followed by data augmentation in which more data is generated from the training set by applying various transformations. By transforming existing images through rotation, cropping, horizontal shifts, vertical shifts, shear, zooming, flipping, reflection, and normalization, more amounts of data are obtained to extract features from the image using various filters in convolution models. Regions such as eyes, lips, and dominant edges were detected. The authors trained the model for 25 epochs and 32 batches to minimize the loss function for each epoch. Physiological signals are extracted from this model with a CNN and fully connected network layers. As soon as the model was trained, the authors gave the data from a web camera as input and checked the corresponding results. For the classification of different human subjects' emotional states using the dataset FER2013, CNN achieved 62% accuracy.

Shen *et al.* used video data to implement a FER system. The features are extracted from the thermal images of facial sub-regions [20]. The Adaboost classifier is applied to categorize the facial expressions in terms of arousal and valence. As a final step, the visible and thermal image facial expression (USTC-NVIE) database was combined to attain the accuracy of 5.3% and 76.7% comparing high and low levels of arousal and valence, respectively. Goulart *et al.* used visible and thermal images to acquire facial expressions in child–robot interaction [21]. The authors included 17 children in the age range from 8 to 12 years in their study. They identified the facial region of interest (ROI) in colour images using the Viola-Jones algorithm. These facial ROIs are transferred to a thermal camera plane by multiplying the homography matrix attained from the camera calibration. According to the experts, to determine the suitable ROI, the error probability was computed using the manually marked reference frame. To infer the emotions, feature extraction, dimensionality reduction, and pattern classification were applied. This ROI selected based on principal component analysis and linear discriminant analysis achieved a mean accuracy of 85.7% and 81.8% respectively for the recognition of five different emotions.

Said *et al.* conducted a study on a face-sensitive convolutional neural network (FS-CNN) to recognize the various emotions [22]. They detected the facial emotions using large-scale images and then anatomical landmarks were analyzed to foresee emotions as expressions. The FS-CNN is comprised of two stages such as patch cropping and CNN. Initially, high-resolution images are scanned for faces, and those faces are then cropped for further processing. CNN was used in the second stage to predict facial expressions using landmarks analytics. The authors obtained a mean precision of 95% for the FS-CNN implemented in the UMD faces dataset.

The proposed work presented in this study demonstrated the feasibility of a system for detecting angry, happy, neutral, and sad emotions in thermal, and digital images using a FER system based on deep neural networks. Furthermore, the deep learning models of emotion recognition in this study were trained by using only 4900 images. It would be more beneficial to use a large collection of digital and thermal images from 50 normal human subjects as inputs for the proposed models. Our next step will be to implement the six-class method for detecting emotions from digital and thermal images, such as angry, joyful, neutral, sad, fearful, and surprised. We currently focus on the classification of emotions in adults 20–30; however, in the future, we will use deep learning to investigate emotions in adults in the age group of above 40 years.

2.6   Conclusion

In the proposed study, deep learning techniques such as MobileNet-V2, and DenseNet-121 are applied to digital and thermal images for the detection of facial emotion. The proposed modified DenseNet-121 provided better accuracy compared to other pre-trained models in the categorization of different facial emotion recognition, namely, angry happy, neutral, and sad using digital and thermal images of normal human subjects. The overall accuracy for DenseNet-121 of digital and thermal images was found to be 94.9% and 95.1%, whereas the MobileNet-V2 of digital and thermal images produced an accuracy of 82.1% and 91.4%. The performance of deep learning techniques using thermal imaging provided better accuracy compared to digital images in facial emotion recognition. It has been shown that the proposed DenseNet-121 model has better discrimination power in recognizing angry, happy, neutral, and sad facial emotions from digital and thermal images. The proposed study showed the relevant demonstration for better performance of thermal imaging compared to the digital imaging method in facial expression detection.

Acknowledgements

The authors would like to convey their heartfelt appreciation to SRM Institute of Science and Technology, Kattankulathur, Chennai, Tamil Nadu, India, for providing the facilities on campus for the data acquisition process.

Funding statement

No specific grant or funding from any public, private, or nonprofit agency is received for this study.

Conflict of interest

All the authors have no conflicts of interest.

Ethical clearance

The Institutional Ethical Committee (SRM Hospital and Research Centre, Kattankulathur, Tamil Nadu, India) has approved this study with approval number 2992/IEC/2021.

Data availability statement

The author, Snekhalatha, can provide the data that support the findings of this study upon request. Due to privacy and ethical concerns, the data is not publicly available.

References

[1]   Tian, Y.I., Kanade, T., and Cohn, J.F. (2001). Recognizing action units for facial expression analysis. *IEEE Transactions on Pattern Analysis and Machine Intelligence* 23(2), 97–115.

[2]   Brzezinski, R.Y., Rabin, N., Lewis, N. *et al.* (2021). Automated processing of thermal imaging to detect COVID-19. *Scientific Reports* 11, 17489. https://doi.org/10.1038/s41598-021-96900-9

[3]   Filippini, C., Perpetuini, D., Cardone, D., Chiarelli, A., and Merla, A. (2020). Thermal infrared imaging-based affective computing and its

application to facilitate human robot interaction: A review. *Applied Sciences* 10, 2924.

[4]   Snekhalatha, U. and Krishnan, P.T. (2021). Automated detection of oro-facial pain from thermograms using machine learning, and deep learning approaches. *Expert Systems* 38(7), 1–14. https://doi.org/10.1111/exsy. 12747

[5]   Khattak, A., Asghar, M.Z., Ali, M., and Batool, U. (2022). An efficient deep learning technique for facial emotion recognition. *Multimedia Tools and Applications* 81, 1649–1683. https://doi.org/10.1007/s11042-021-11298

[6]   Mehendale, N. (2020). Facial emotion recognition using convolutional neural networks (FERC). *SN Applied Sciences* 2, 446. https://doi.org/ 10.1007/s42452-020-2234-1

[7]   Wahyono, D., Ashar, M., Fadlika, I., Asfani, K., Saryono, D., and Sunarti. A new computational intelligence for face emotional detection in ubiqui-tous, in *2019 International Conference on Electrical, Electronics, and Information Engineering (ICEEIE)*, 2019, pp. 148–153. https://doi.org/ 10.1109/ICEEIE47180.2019.8981420

[8]   Mollahosseini, A., Chan, D., and Mahoor, M.H. Going deeper in facial expression recognition using deep neural networks, in *2016 IEEE Winter Conference on Applications of Computer Vision (WACV)*, 2016, pp. 1–10. doi:10.1109/WACV.2016.7477450.

[9]   Mukhopadhyay, M., Dey Shaw, A.R.N., and Ghosh, A. Facial emotion recognition based on Textural pattern, and Convolutional Neural Network, in *2021 IEEE 4th International Conference on Computing, Power, and Com-munication Technologies (GUCON)*, 2021, pp. 1–6. https://doi.org/10.1109/ GUCON50781.2021.9573860.

[10]   Kola, D.G.R. and Samayamantula, S.K. (2021). A novel approach for facial expression recognition using local binary pattern with adaptive window. *Multimedia Tools Applications* 80, 2243–2262. https://doi.org/10.1007/ s11042-020-09663-2

[11]   Bhattacharyya, A., Chatterjee, S., Sen, S., Sinitca, A., Kaplun, D., and Sarkar, R. (2021). A deep learning model for classifying human facial expressions from infrared thermal images. *Scientific Reports* 11(1), 20696. https://doi.org/10.1038/s41598-021-99998-z

[12]   Elbarawy, Y.M., Ghali, N.I., and El-Sayed, R.S. (2019). Facial expressions recognition in thermal images based on deep learning techniques. *Interna-tional Journal of Image, Graphics, and Signal Processing (IJIGSP)* 11(10), 1–7. https://doi.org/10.5815/ijigsp.2019.10.01

[13]   Seidaliyeva, U., Akhmetov, D., Ilipbayeva, L., and Matson, E.T. (2020). Real-time, and accurate drone detection in a video with a static background. *Sensors* 20(14), 3856.

[14]   Dertat, A. Applied Deep Learning – Part 4: Convolutional Neural Networks, 2017. https://towardsdatascience.com/applied-deep-learning-part-4-convolu-tional-neural-networks-584bc134c1e2.

[15]   Howard, A.G., Zhu, M., Chen, B., *et al.* MobileNets: Efficient Convolutional Neural Networks for Mobile Vision Applications. *arXiv preprint arXiv:1704. 04861*, 2017.

[16]   Tsang, S.H. Review: MobileNetV2-Lightweight Model (Image Classification), 2019.

[17]   Li, Y., Zeng, J., Shan, S., and Chen, X. (2018). Occlusion aware facial expression recognition using CNN with attention mechanism. *IEEE Transactions on Image Processing* 28(5), 2439–2450.

[18]   Bodavarapu, P.N.R. and Srinivas, P.S. (2021). An optimized neural network model for facial expression recognition over traditional deep neural networks. *International Journal of Advanced Computer Science and Applications* 12(7), 443–451.

[19]   Shirisha, K. and Buddha, M. (2020). Facial emotion detection using convolutional neural network. *International Journal of Scientific & Engineering Research* 11(3), 51.

[20]   Shen, P., Wang, S., and Liu, Z. Facial expression recognition from infrared thermal videos. *Intelligent Autonomous Systems*, vol. 12, 2013, Springer: Berlin, pp. 323–333.

[21]   Goulart, C., Valadão, C., Delisle-Rodriguez, D., Funayama, D., Favarato, A., Baldo, G., Binotte, V., Caldeira, E., and Bastos-Filho, T. (2019). Visual, and thermal image processing for facial specific landmark detection to infer emotions in a child–robot interaction. *Sensors* 19(13), 2844.

[22]   Said, Y. and Barr, M. (2021). Human emotion recognition based on facial expressions via deep learning on high-resolution images. *Multimedia Tools and Applications* 80(16), 25241–25253.

Chapter 3

Gender and emotion recognition from EEG and eye movement patterns

*Maneesh Bilalpur[1], Seyed Mostafa Kia[2],
Mohan Kankanhalli[3], Ramanathan Subramanian[4] and
M. Murugappan[5,6]*

Emotion recognition (ER) and gender recognition (GR) through non-invasive sensors are highly useful in the assessment of psychological and physiological behavior. The purpose of this chapter is to examine whether the implicit behavioral cues found in electroencephalogram (EEG) signals as well as eye movements can be used to recognize gender (GR) and emotion (ER) from psychophysical behavior. The cues examined are obtained using inexpensive, off-the-shelf sensors. There were 28 users (14 males) who recognized Ekman's basic emotions from unoccluded faces (no mask) and partially occluded faces (eye or mouth masks); EEG responses encoded gender-specific differences, while eye movements were indicative of the perception of facial emotions. The use of convolutional neural networks and AdaBoost for classification demonstrates (a) that with EEG and eye characteristics, reliable GR (peak area under the ROC curve (AUC) of 0.97) and ER (peak AUC of 0.99) are feasible, (b) females exhibit differential cognitive processing of negative emotions based on event-related potential patterns, and (c) gender differences in eye gaze are observed under partial face occlusions, such as eye and mouth masks.

Keywords: Gender and emotion recognition; Emotional face perception; Implicit user behavior; Electroencephalography; Eye gaze tracking; Unoccluded and occluded faces

[1]School of Computing and Information, University of Pittsburgh, USA
[2]Tilburg School of Humanities and Digital Sciences, Department of Cognitive Science and Artificial Intelligence, Tilburg University, The Netherlands
[3]School of Computing, National University of Singapore, Singapore
[4]School of Information Technology and Systems, University of Canberra, Australia
[5]Intelligent Signal Processing (ISP) Research Lab, Department of Electronics and Communication Engineering, Kuwait College of Science and Technology, Kuwait
[6]Department of Electronics and Communication Engineering, Faculty of Engineering, Vels Institute of Sciences, Technology, and Advanced Studies, India

3.1   Introduction

Gender human–computer interaction (HCI) [1] and affective HCI [2] have evolved as critical HCI sub-fields, as it is critical for computers to appreciate and adapt to the user's gender and emotional state. Inferring users' soft biometrics such as gender and emotion would benefit interactive and gaming systems in terms of (a) visual and interface design [3,4], (b) game and product recommendation (via ads) [5,6], and (c) provision of appropriate motivation and feedback for optimizing user experience [7]. Gender recognition (GR) and emotion recognition (ER) systems primarily work with facial [8,9] or speech [10,11] cues which are biometrics encoding a person's identity. Also, they can be recorded without the user's knowledge, posing grave privacy concerns [12]. This work examines GR and perceived ER (emotion of the presented stimulus) from implicit user behavioral signals, in the form of electroencephalogram (EEG) brain signals and eye movements. Implicit behavioral signals are inconspicuous to the outside world and cannot be recorded without express user cooperation making them privacy-compliant [13]. Also, behavioral signals such as EEG and eye movements are primarily anonymous as little is known regarding their uniqueness to a person's identity [14].

Specifically, we attempt GR and perceived ER using signals captured by commercial, off-the-shelf devices that are minimally intrusive, affordable, and popularly used in gaming as input or feedback modalities [15,16]. The Emotiv EEG wireless headset consists of 14 dry (plus two reference) electrodes having a configuration as shown in Figure 3.1. While being lightweight, wearable, and easy to use, neuro-analysis with Emotiv can be challenging due to relatively poor signal quality. Likewise, EyeTribe is a low-cost eye tracker whose suitability for research has been endorsed [17]. We show how relevant gender and emotion-specific information is captured by these low-cost devices via examination of event-related potential (ERP) and eye fixation patterns, and through recognition experiments.

We set out to discover gender differences in human visual perception by designing a facial emotion recognition (FER) experiment. Males and females respond differently to affective information [7,18–20], and user eye movements are also characteristic of the perceived facial emotion [21–25], enabling stimulus ER. Our study performed with 28 viewers (14 males) confirms that women achieve superior FER, mirroring prior findings. Hypothesizing that enhanced female emotional sensitivity should reflect via their implicit behavior, we examined EEG and eye-gaze patterns to find that (1) stronger ERPs are observed for females while processing negative facial emotions, (2) female eye-gaze is more focused on the eyes for the purpose of FER, and (3) emotion and information-specific gender differences manifest starkly, enabling better GR under particular stimulus conditions.

Building on preliminary results [20,26], our work makes the following research contributions: (a) While prior works have identified gender differences in emotional behavior, this is one of the first works to expressly perform GR and perceived ER from implicit behavioral signals. (b) Apart from recognition experiments, we show that the employed devices capture meaningful perceptual information, as typified by gender and emotion-specific ERPs, and the extent of fixation

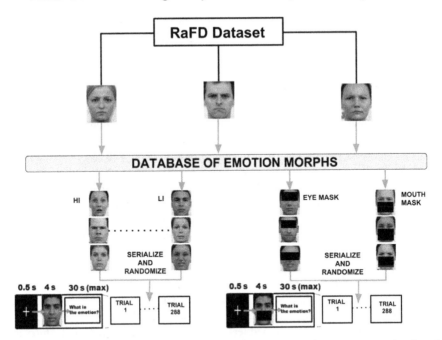

Figure 3.1   *Experimental protocol: viewers were required to recognize the facial emotion from either an unmasked face (Session 1) or from an eye/ mouth-masked face (Session 2). Trial timelines for the two conditions are shown at the bottom.*

over the eyes. (c) We demonstrate a significant performance improvement in GR with deep learning and boosting methods over traditional approaches presented in [20,26]. (d) The use of minimally intrusive, off-the-shelf, and low-cost devices affirms the ecological validity of our study and the utility of our experimental design for large-scale user profiling.

This chapter is organized as follows: Section 3.2 reviews related works. Section 3.3 describes our study methodology. Section 3.4 examines explicit user responses, which are correlated with implicit behaviors in Section 3.5, followed by GR and ER experiments. Section 3.6 summarizes and concludes the chapter.

3.2   Related works

Many works perform ER with implicit behavioral cues such as eye movements, EEG, electromyogram (EMG), galvanic skin response (GSR) [27–31], etc. However, very few works estimate soft biometrics such as gender, cognitive load, and personality traits with such signals [32–34]. Also, while some works isolate emotion and gender differences in eye movement and EEG responses to emotional faces [24,25,35,36], these differential features are never utilized for gender prediction. To position our work concerning the literature, this section reviews related work on (a) user-centered ER and (b) gender differences in emotional face processing.

3.2.1   User-centered ER

Emotions evoked by multimedia stimuli are predicted via content-centered or user-centered methods. Content-centered methods attempt to find emotional multimodal features [37–40], while user-centered methods monitor user behavioral cues (eye movements, EEG signals, etc.) to deduce the emotion perceived by or evoked in the user. As emotions are subjective, many user-centered approaches predict emotions by examining both explicit and implicit user behavioral cues. Conspicuous facial cues are studied to detect multimedia highlights in [9], while physiological measurements are utilized to model emotions induced by music and movie scenes [29,30]. EEG and eye movements are two popular modalities employed for ER, and many works have used a combination of both [20,27,28] or signals exclusively [21,24,29,31,36,41,42].

Valence (positive vs negative emotion) recognition from eye-gaze features is achieved in [21]. ER from EEG and pupillary responses is discussed in [27]. Deep unsupervised ER from raw EEG data is proposed in [41], and its effectiveness is shown to be comparable to designed features. Differential entropy EEG features are extracted to train an integrated deep belief network plus a hidden Markov model for ER [42]. A differential auto-encoder that learns shared representations from EEG and eye-based features is proposed for valence recognition (VR) in [28]. Almost all these works employ lab-grade eye trackers and EEG sensors which are bulky and intrusive, and therefore preclude naturalistic user behavior. In contrast, we employ low-cost and off-the-shelf sensors, which enable users to react naturally to the presented stimuli and allow for user profiling at scale.

3.2.2   Gender differences in ER

As facial emotions denote critical non-verbal communication cues in social interactions, many psychology studies have studied human FER. Certain facial features encode emotions better than others; the eyes, nose, and mouth are the most attractive facial regions [43,44]. Visual attention is localized around the eyes for mildly emotive faces, but the nose and mouth attract substantial eye fixations in highly emotive faces [45]. An eye-tracking study [25] notes that distinct eye fixation patterns emerge for different facial emotions. The mouth is the most informative for joy and disgust emotions, whereas the eyes mainly encode information relating to sadness, fear, anger, and shame. A similar study [46] notes more fixations on the upper face half for anger as compared to disgust, while no differences are observed on the lower face half for the two emotions. However, humans may find it difficult to distinguish between similar facial emotions—examples are the high overlap rate between the fear–surprise and anger–disgust emotion pairs [44,47].

Multiple works have discovered gender differences during facial emotion processing. Females are generally better at FER irrespective of age [48]. Other FER studies [18,19,49] also note that females recognize facial emotions more accurately than males, even under partial information. Some evidence also points to females achieving faster FER than males [50,51]. Gender differences in gaze patterns and neural activations have been found while viewing emotional faces; females' tendency to fixate on the eyes positively correlates with their ER capabilities, while men tend to look at the mouth for emotional cues [48,50]. Likewise, EEG ERPs reveal that negative facial emotions are processed differently and rapidly by women, and do not necessarily entail selective attention toward emotional cues [35,52].

An exhaustive review of GR methodologies is presented in [32], and the authors evaluate GR methods using metrics like universality, distinctiveness, permanence, and collectability. While crediting bio-signals like EEG and electrocardiography for their accuracy and trustworthiness, the authors also highlight the invasiveness of biosensors. The sensors used in this work are minimally intrusive, enabling a naturalistic user experience, while also recording meaningful emotion and gender-related information. Among user-centric GR works, EEG and speech features are proposed for age and GR in [53]. Recent behavioral studies such as [54,55] examine gender differences in traumatic brain injury participants during a working memory task. These studies observe significant gender-based differences between brain-injured and healthy participants.

3.2.3 Analysis of related work

Close examination of the literature reveals (1) many works achieve evoked ER from user-centered cues, both conspicuous and latent, and a handful have discovered gender differences in gaze patterns and neural activations; nevertheless, very few works expressly predict gender from implicit user cues. Differently, we employ implicit signals for GR and perceived ER, and achieve reliable gender and valence detection (area under the ROC curve (AUC) > 0.9); (2) Our GR/perceived ER features are acquired from low-cost, off-the-shelf sensors, which record inferior user signals even while enabling natural user behavior. We nevertheless show how these sensors capture meaningful information via the analysis of ERPs and fixation distribution patterns; (3) Different to prior works which either analyze explicit or implicit user responses to discover gender differences or do not expressly isolate bio-signal patterns, in contrast, we show multiple similarities among explicit and implicit user behaviors to validate our findings.

3.3   Materials and methods

Our study objective was to examine user behavior while viewing unoccluded/partly occluded emotional faces and predict user gender therefrom. We hypothesized that gender differences would be captured via EEG and eye-gaze patterns. Also, eye movements are known to be characteristic of the perceived facial emotion [25], which enables inference of the stimulus emotion; we limit ourselves to predicting the stimulus valence, i.e., whether the face presented to the viewer exhibits a positive or negative emotion.

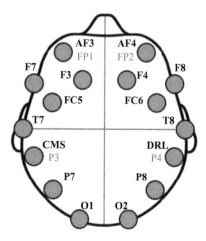

*Figure 3.2   Emotiv Epoc+ electrode configuration: The headset comprises
14 sensors plus two reference electrodes (CMS and DRL). The
figure is adapted from under Creative Commons License CC BY-3.0.*

We designed a study to examine gender differences in visual emotional face processing with (a) fully visible faces, and (b) faces with the eye and mouth regions occluded via a rectangular mask (Figure 3.2). The salience of the occluded features toward conveying facial emotions is well-known [25,45]. Specifically, we considered emotional faces corresponding to four conditions: exhibiting high-intensity (HI) and low-intensity (LI) emotions, and additionally, HI emotions upon occluding the eye (eye mask) or mouth (mouth mask) regions (Figure 3.1). Since we hypothesized that emotion perception under facial occlusion would be considerably difficult (cf. Table 3.2), we did not study masked and mildly emotive faces.

(a) **Participants:** Twenty-eight students from different nationalities (14 males, age 26.1 ± 7.3 and 14 females, age 25.5 ± 6), with normal or corrected vision, took part in our study. All users provided informed consent and were presented with a token fee for participation as directed by the ethics committee.

(b) **Stimuli:** We used the emotional faces of 24 models (12 males, 12 females) from the Radboud Faces Database (RaFD) [57]. RaFD includes the facial emotions of 49 models rated for clarity, genuineness, and intensity, and the 24 models were chosen such that their Ekman facial emotions (Anger (A), Disgust (D), Fear (F), Happy (H), Sad (Sa) and Surprise (S)) were roughly matched based on these ratings. We then morphed the emotive faces from neutral (0% intensity) to maximum (100% intensity) to generate intermediate morphs in steps of 5%. Derived morphs with 55–100% intensity were used as HI emotions, and 25–50% were used as LI emotions. Eye and mouth-masked faces were automatically generated upon locating facial landmarks via Openface [58] over the HI morphs. The eye mask covered the eyes and nasion, while the mouth mask covered the mouth and the nose ridge. All stimuli were resized to 361 × 451 pixels, encompassing a visual angle of 9.1° and 11.4° about x and y at a 60 cm screen distance.

(c) **Protocol:** The experimental protocol is outlined in Figure 3.1 and involves the presentation of unmasked and masked faces to viewers over two separate sessions, with a break in between to avoid fatigue. We chose one face per model and emotion, resulting in 144 face images (1 morph/emotion × 6 emotions × 24 models). In the first session (no-mask condition), these faces were shown in random order and were again re-presented randomly with an eye or mouth mask in the second session. We ensured a 50% split of the HI and LI morphs in the first session, and eye/mouth-masked faces in the second.

During each trial, an emotional face was displayed for 4 s preceded by a fixation cross for 500 ms. The viewer then had a maximum of 30 s to make one out of seven choices concerning the facial emotion (six Ekman emotions plus neutral) via a radio button. Neutral faces were only utilized for morphing purposes and were not used in the experiment. Viewers' EEG signals were acquired via the 14-channel Emotiv Epoc+ device (Figure 3.2), and eye movements were recorded with the Eyetribe tracker during the trials. The face-viewing experiment was split into four segments to facilitate sensor re-calibration and minimize data recording errors and took about 90 min to complete.

3.3.1   User responses

We first compare male and female sensitivity to emotions based on explicitly observed user response times (RTs) and recognition rates (RRs) and will then proceed to examine their implicit eye movement and EEG responses. RT is defined as the time taken for the participant to recognize the stimulus emotion from the presentation time. The correctness of the participant's response to the stimuli concerning the ground truth emotion is employed to compute the RR. Our experimental design involved four stimulus types (HI, LI, eye mask, and mouth mask), and two user types (male and female), resulting in 4 × 2 factor conditions.

3.3.1.1   Response time

Overall user RTs for the HI, LI, eye mask, and mouth mask conditions were, respectively, found to be 1.44 ± 0.24, 1.52 ± 0.05, 1.17 ± 0.12, and 1.25 ± 0.09 s, implying that FER was instantaneous, and viewer responses were surprisingly faster with masked faces. Fine-grained comparison of male (m) and female (f) RTs across stimulus types (Figure 3.3) revealed that females (μRT = 1.40 ± 0.10 s) were generally faster than males (μRT = 1.60 ± 0.10 s) at recognizing HI emotions. There was no significant difference in RTs for LI emotions. Female alacrity nevertheless decreased for masked faces, with males responding marginally faster for eye-masked (μRT(m) = 1.13 ± 0.11 s vs μRT(f) = 1.21 ± 0.13 s), and both genders responding with similar speed for mouth-masked faces (μRT(m) = 1.24 ± 0.10 s vs μRT(f) = 1.25 ± 0.09 s).

Response Time

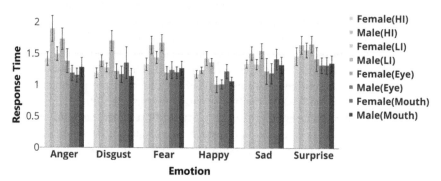

Figure 3.3   Emotion-wise RTs of females and males in various conditions. y-axis denotes RT in seconds. Error bars denote unit standard error (best viewed in color).

Recognition Rate

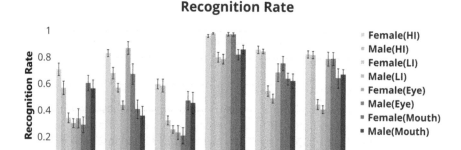

Figure 3.4   Emotion-wise RRs in various conditions. Error bars denote unit standard error.

3.3.1.2   Recognition rates

While females recognized facial emotions marginally faster, we examined if they also achieved superior FER. Overall, RRs for unoccluded HI emotions (μRR = 77.6) were expectedly higher than for eye-masked (μRR = 59.7), mouth-masked (μRR = 63.5), and LI emotions (μRR = 49.1). Happy faces were recognized most accurately in all four conditions. Specifically focusing on gender differences (Figure 3.4), females recognized facial emotions more accurately than males and this was particularly true for negative (A, D, F, Sa) emotions; male vs female RRs for these emotions differed significantly in the HI (μRR(m) = 54.3 vs μRR(f) = 61.2, $p < 0.05$) and eye mask conditions (μRR(m) = 51.8 vs μRR(f) = 58.1, $p < 0.05$), and marginally for the mouth mask condition (μRR(m) = 52 vs μRR(f) = 55.8, $p = 0.08$). Males marginally outperformed females in the LI condition (μRR(m) = 49.8 vs μRR(f) = 47.7). Overall, (a) HI emotion morphs were recognized more accurately than LI morphs, (b) females recognized negative emotions better, (c) happy was the easiest emotion to recognize, and (d) higher RRs (across gender) were noted in the eye mask condition for four of the six Ekman emotions, implying that deformations around the mouth were more informative for FER under occlusion.

3.3.2   Analyzing implicit responses

As females achieved quicker and superior FER for negative emotions, we hypothesized that these behavioral differences should also be reflected via implicit eye-gaze and EEG patterns. We first describe the EEG and eye movement descriptors employed for analyses, before discussing (stimulus) ER and (user) GR results.

3.3.2.1   EEG pre-processing

We extracted EEG epochs for each trial (4.5 s of stimulus-plus-fixation viewing time at a 128 Hz sampling rate), and the 64 leading pre-stimulus samples were used to remove the DC offset. This was followed by (a) EEG band-limiting to within 0.1–45 Hz using a windowed *sinc* FIR filter, (b) removal of noisy epochs via visual inspection, and (c) independent component analysis (ICA) to remove artifacts relating to eye blinks, and eye/muscle movements. Muscle movement artifacts in EEG are mainly concentrated in the 40–100 Hz band and are removed upon band-limiting and via inspection of ICA components. Finally, a 7168-dimensional (14-channel × 4 s × 128 Hz) EEG feature vector was generated and fed to different classifiers for GR and ER.

3.3.2.2   Event-related potentials

ERPs are time-locked neural responses related to sensory and cognitive events and denote the EEG response averaged over multiple users and trials. For example, P300 and N100, N400 are exemplar ERPs which are typically noted around 300, 100, and 400 ms post-stimulus onset. ERPs occurring within 100 ms post-stimulus onset are stimulus-related (exogenous), while later ERPs are cognition-related (endogenous). We examined the leading 128 EEG epoch samples (one second of data) for ERP patterns relating to emotion and gender.

Prior works have observed ERP-based gender differences from lab-grade sensor recordings [35,52,59]. Specifically, [52] notes enhanced negative ERPs for females in response to negative valence stimuli. However, capturing ERPs with commercial devices is challenging due to their low signal-to-noise ratio [60]. Figures 3.5 and 3.6 present the P300, visual N100, and N400 ERP components in the occipital O1 and/or O2 electrodes (see Figure 3.1 for sensor positions) corresponding to various face morphs. Note that the occipital lobe is the visual processing center in the brain, as it contains the primary visual cortex.

Comparing O1/O2 male and female ERPs for positive (H, Su) and negative (A, D, F, Sa) emotions, no significant differences can be observed between male positive and negative ERP peaks for HI or LI faces (columns 3 and 4 in Figure 3.5).

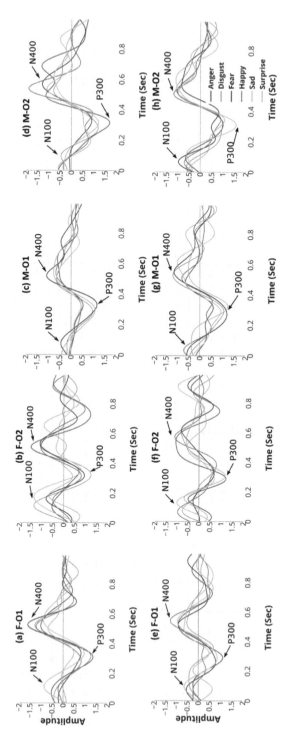

Figure 3.5   ERPs for HI morphs (top) and LI morphs (bottom): (left to right) O1 and O2 ERPs for females and males. The y-axis shows ERP amplitude in μV, refer to (h) for legend. As per convention, ERPs are plotted upside down (view in color).

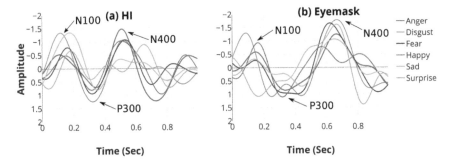

Figure 3.6   Female ERPs in the (left) HI and (right) eye mask conditions from the O2 channel. Best viewed in color.

However, we observe stronger N100 and P300 peaks in the negative female ERPs for both HI and LI faces (columns 1 and 2). Also, a stronger female N400 peak can be noted for HI faces consistent with prior findings [52]. Contrastingly, lower male N100 and P300 latencies are observed for positive HI emotions, with the pattern being more obvious at O2. Likewise, lower male N400 latencies can be generally noted at O2 for positive emotions. The positive vs negative ERP difference for females is narrower for LI faces, revealing the difficulty in identifying mild LI emotions. That LI faces produce weaker ERPs at O1 and O2 than HI faces further supports this observation.

Figure 3.6 shows female ERPs observed in the occipital O2 electrode for the HI and eye mask conditions. Clearly, one can note enhanced N100 and P300 ERP components for negative HI emotions (Figure 3.6 (left)). This effect is attenuated in the eye mask (Figure 3.6 (right)) and mouth mask cases, although one can note stronger N400 amplitudes for D and F with the eye mask. This ERP pattern is invisible for males, confirming their gender specificity. Overall, ERP patterns affirm that gender differences in emotional face processing can be reliably isolated with the low-cost Emotiv device. We performed ERP analysis for all electrodes, but the ERP analysis primarily identified patterns in the occipital lobes (these are also task-relevant as described). While it is possible that more ERPs could be found in the proximity of the occipital lobe, our ERP analyses confirm that gender-based emotion perception differences can be identified even using low-cost, off-the-shelf sensors such as the Emotiv headset.

3.3.2.3   Eye-tracking analysis

Gender differences in gaze patterns during emotional face processing have been noted by prior works [25,48]. We used the low-cost Eyetribe1 device with 30 Hz sampling to record eye movements. Raw gaze data output by the tracker was processed to compute fixations (prolonged gazing at scene regions to assimilate visual information) and saccades (transition from one fixation to another) via the EyeMMV toolbox [61]. Upon extracting fixations and saccades, we extracted features employed for VR in [21] to compute an 825-dimensional feature vector

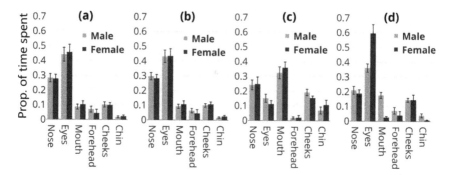

Figure 3.7   (a–d) Fixation duration distributions for males and females in the HI, LI, eye, and mouth mask conditions. Error bars denote unit standard deviation.

including saccade velocity, saccade length, fixation duration, etc. For the exhaustive list of features for our analyses. These features include static and dynamic attributes of saccades and fixations, and their definitions refer to [21].

3.3.2.4 Fixation analysis

To study gender differences in fixating patterns, we computed the distribution of fixation duration (FD) over six facial regions, namely, eyes, nose, mouth, cheeks, forehead, and chin. Figure 3.7 presents the male and female FD distribution over these facial regions for various conditions. For both genders, the time spent examining the eyes, nose, and mouth accounted for over 80% of the total FD, with eyes (≈45%) and nose (≈30%) attracting maximum attention as observed in [25,36,45]. Relatively similar FD distributions were noted for both genders with HI and LI morphs (Figure 3.7(a) and (b)). Figure 3.7(c) and (d) presents FD distributions in the eye and mouth mask conditions. An independent t-test revealed a significant difference ($p < 0.05$) between male and female FDs for the eye region in the mouth mask condition; prior works [23] have observed that females primarily look at the eyes for emotional cues, which is mirrored by longer FDs around the eyes in the HI and mouth mask conditions. Figure 3.7(c) shows that when eye information is unavailable, females tend to focus on the mouth and nasal regions for FER.

3.4 Experiment and results

Having noticed gender-specific patterns in user responses, EEG ERPs, and eye movements, we attempted binary perceived ER and GR employing EEG features, eye-based features, and their combination for the various conditions. Recognition was attempted only on trials where viewers correctly recognized the presented facial emotion. We considered (i) EEG features, (ii) eye-based features, (iii) concatenation of the two (early fusion or EF), and (iv) probabilistic fusion of the EEG and eye-based classifier outputs (late fusion or LF) for our analyses. The West technique [62] was used to fuse the EEG and eye-based outputs and denotes maximum possible recognition performance as optimal weights maximizing the test AUC metric were determined via a 2D grid search.

We considered the AUC (plotting true vs false positive rates, as the performance metric for benchmarking). AUC is suitable for evaluating classifier performance on imbalanced data (as our ER task involves classifying four negative emotions vs two positive emotions coupled with the differences in RRs as shown in Figure 3.4), and a random classifier will achieve an AUC of 0.5 [63]. As we attempted recognition with few training data, we reported ER/GR results over five repetitions of 10-fold cross-validation (CV) (i.e., a total of 50 runs). CV is typically used to overcome the overfitting problem and train generalizable classifiers on small datasets.

3.4.1   Baseline classification approaches

As baselines, we considered the naive Bayes (NB), linear support vector machine (LSVM), and radial-basis SVM (RSVM) classifiers. NB is a generative classifier that estimates the test label based on the maximum a posteriori criterion, $p(C|X)$, assuming class-conditional feature independence. C and X, respectively, denote the test class label and feature vector. LSVM and RSVM denote the linear and radial basis kernel versions of the SVM classifier. SVM hyperparameters C (LSVM) and γ (RSVM) were tuned from within $[10^{-4}, 10^4]$ via an inner 10-fold CV on the training set.

3.4.2   NN for EEG-based GR and ER

Deep learning frameworks have recently become popular due to their ability to automatically learn optimal task features from raw data, thereby obviating the need for data cleaning and feature extraction. However, unlike in image or video processing, temporal dynamics of the human brain are largely unclear; designing relevant features is therefore hard. We hypothesized that convolutional neural networks (CNNs) would encode EEG patterns efficiently and be robust to artifacts. These factors inspired us to feed raw EEG data to a CNN, without any data processing; pre-processed EEG data was nevertheless fed to the above baseline classifiers.

The terminology explained below is limited to the CNN model used in this work. With deep neural network research rapidly evolving, new layers such as unpooling and attention have come into existence.

1. **Receptive field:** Receptive field of a kernel refers to the portion of the input signal that a kernel can access.
2. **1D convolution:** A one-dimensional (1D) convolution layer performs traditional convolution (weighted sum of the input signal and kernel) using a 1D convolution kernel.
3. **Pooling:** For a given kernel size, the pooling layer performs a simple operation on the input receptive field such as average (for average pooling) or max (max pooling). Pooling reduces the dimensionality of the feature space.

4. **Dropout:** Dropout helps in training the CNN model by introducing regularization into the architecture. The input to a dropout layer is dropped (set to zero) with a probability of p.
5. **Batch normalization:** As the number of layers increases, the distributional shift in the intermediate layer features of the NN increases. Batch normalization helps in preserving the original distribution. It helps in training very deep models by acting as a regularizer.
6. **Fully connected (FC) layer:** Unlike convolution layers that have a receptive field limited by their kernel size (e.g., a 1×3 kernel has a field of three samples of the input even if the input is longer), the FC layers are fully connected and hence the output is the weighted sum of all inputs.
7. **Softmax:** The softmax layer takes a multinomial input to scale them to probabilities in the 0 to 1 range. This is typically used in neural networks to map the pre-softmax final layer outputs (referred to as logits) to probabilities so that likelihood-based loss functions such as cross-entropy can be used for training.

We adopted a three-layer CNN [64] to learn a robust EEG representation for gender and VR. Three convolution layers together with rectified linear unit activation and average pooling layers are stacked to learn EEG descriptors (Figure 3.8). The convolutions employed are one-dimensional over time. Batch normalization [65] is used after the third CNN layer to minimize covariate shift and accelerate training. To prevent overfitting, we used dropout after the FC layer with 128 neurons. A softmax over two output neurons was used for classification. The choice of architecture is based on a series of experiments varying the number of layers and hyperparameters (Table 3.1) aimed at balancing the performance-overfitting tradeoff. Experiments revealed that the three-layer CNN offered optimal performance without overfitting on the training set.

The number of kernels increases with network depth (cf. Figure 3.8) analogous to the Visual Geometry Group architecture [66]. We optimized the network for categorical cross-entropy loss using stochastic gradient descent with Nesterov momentum and weight decay. To decrease data dimensionality and avoid overfitting without sacrificing temporal EEG dependencies, we downsampled EEG data

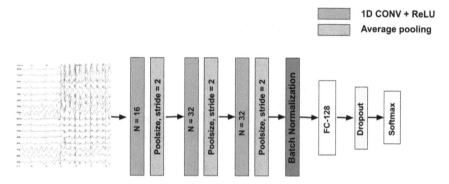

Figure 3.8   CNN architecture showing various layers in the model and parameters

Table 3.1   Summary of the deep CNN model

Hyperparameter	Value
Learning rate	0.01
Kernal size	3×3
Stride factor	2
Pool size	2
Batch size	32
#Kernal (layer wise)	16, 32, 64
Momentum factor	0.9
Weight decay	0.0001
Dropout	0.1

to 32 Hz (reducing per-channel EEG dimensionality to $32 \times 4 = 128$). A tenth of the training data was used for validation and early stopping was enforced to prevent overfitting. CNN hyperparameters specified in Table 3.1 were either adopted from [64] or upon cross-validation.

CNN was traditionally trained for GR with normal initializers (normally distributed weights). Conversely, we adopted two-stage training for ER. In the first stage, the model was pre-trained with EEG data acquired over all four experimental conditions (Figure 3.2); the second stage involved fine-tuning with data for a specific condition. The objective here was to extract valence-related features irrespective of stimulus type in the first stage and fine-tune them to learn condition-specific descriptors. Experiments were designed using Keras [67] with Tensorflow back-end on a 16 GB CPU machine and an Intel i5 processor.

3.4.3   AdaBoost for GR and ER from eye-gaze

We employed the AdaBoost classifier for GR and ER from eye-gaze features. AdaBoost is an ensemble classifier popularly used for face detection [68] and FER [69]. AdaBoost combines several weak classifiers (decision stumps), each marginally better than random, to cumulatively achieve optimal classification. The prediction is based on the weighted output of each weak classifier. Our features and classifier design were inspired by [69], who capture local dependencies via histograms of gaze measures like fixations, saccades, and saliency. We employed Stagewise Additive Modeling using a Multi-class Exponential (SAMME) loss function [70] for training. User data compiled for this study along with the CNN and AdaBoost models employed for GR and ER, respectively, are available at https://github.com/bmaneesh/emotion-xavier.

3.4.4   Experimental results

3.4.4.1   Emotion recognition

We modeled perceived ER as a binary classification problem, where the objective is to categorize positive (H, Su) and negative (A, D, F, Sa) valence stimuli employing gaze and EEG-based cues. ER results with (CNN-based) EEG,

(AdaBoost-based) eye-gaze features, and late fusion of two modalities, with data acquired for different conditions characterized by user-gender (M/F) and stimulus type (HI/LI/eye mask/mouth mask) are shown in Figure 3.9. Evidently, gaze features perform significantly better than EEG. Gaze features achieve near-ceiling VR barring the case where female users view LI emotions. Optimal performance is noted with male data for HI emotions (AUC = 0.99) similar to [71]; ER from female eye-gaze data on HI emotions (AUC = 0.98) is also high.

On the other hand, ER results with EEG largely produced near-chance performance, with only male and female data for the mouth-masked condition being exceptions. These results reveal that the raw EEG features are not optimal for ER. Given the vast difference in ER performance with the gaze and EEG modalities, late fusion results are only occasionally superior to unimodal ones. Fusion slightly outperforms unimodal methods for the eye mask condition for both males (0.98 vs 0.99) and females (0.97 vs 0.98), and noticeably with males for mouth mask faces (0.91 vs 0.94).

Overall, while recognizing the stimulus facial emotion was not our primary objective, the extracted gaze features are nevertheless highly effective for VR. AdaBoost employing gaze features significantly outperforms the CNN trained with EEG data. The ability of eye movements to characterize emotional differences is unsurprising. Distinctive eye movement patterns have been observed while scanning scenes and faces with different emotions [21,24,25]. Eye-based features are found to achieve 52.5% VR accuracy in [21], where emotions are induced by presenting diverse emotional scenes, while our study is specific to emotive faces. Prior neural studies [29,59] that perform ER with lab-grade sensors achieve around 60% VR. The superior performance of eye-based features can be attributed to the fact that the saccade and saliency-based statistics can capture discrepancies in gazing patterns for positive and negative facial emotions [21,24]; in contrast, neural features have only been moderately effective at decoding emotions conveyed by audio-visual music and movie stimuli [29,30,59], which can presumably elicit emotions more effectively than plain imagery.

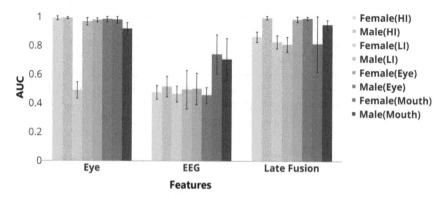

Figure 3.9   VR results for different conditions. Error bars denote unit standard deviation.

3.4.4.2   Gender recognition

GR involves labeling the test EEG or eye-gaze sample as arising from a male/ female user. Table 3.2 presents GR achieved with baseline classifiers, while Table 3.3 shows AUC scores obtained with the EEG-based CNN, and the AdaBoost ensemble fed with eye-gaze features. Both tables present GR results when user data correspond to all emotional faces, and data for each of the six Ekman emotions

Table 3.2   GR results with baseline classifiers over different modalities and their fusion

	AUC		HI	LI	Eye mask	Mouth mask
Emotion wise	ALL	EEG (NB	**0.714 ± 0.002**	**0.600 ± 0.005**	**0.690 ± 0.03**	**0.654 ± 0.04**
		EYE (NB)	0.493 ± 0.013	0.481 ± 0.005	0.471 ± 0.06	0.525 ± 0.05
		Early fusion (RSVM)	0.522 ± 0.035	0.523 ± 0.035	0.540 ± 0.05	0.610 ± 0.07
		Late Fusion (LSVM)	0.549 ± 0.022	0.523 ± 0.035	0.540 ± 0.05	0.610 ± 0.07
	EEG (NB)	A	**0.708 ± 0.064**	0.580 ± 0.74	0.610 ± 0.16	**0.672 ± 0.07**
		D	0.673 ± 0.055	**0.696 ± 0.062**	0.592 ± 0.06	0.650 ± 0.14
		F	0.643 ± 0.059	0.596 ± 0.089	0.605 ± 0.14	0.564 ± 0.12
		H	0.696 ± 0.047	0.668 ± 0.046	0.586 ± 0.07	0.624 ± 0.08
		Sa	0.674 ± 0.048	0.634 ± 0.064	**0.652 ± 0.08**	0.590 ± 0.08
		Su	0.692 ± 0.048	0.636 ± 0.071	0.633 ± 0.08	0.650 ± 0.09
	EYE (NB)	A	**0.601 ± 0.021**	0.565 ± 0.031	0.470 ± 0.20	**0.590 ± 0.16**
		D	0.577 ± 0.011	**0.632 ± 0.009**	0.480 ± 0.14	0.521 ± 0.27
		F	0.595 ± 0.015	0.535 ± 0.029	**0.680 ± 0.25**	0.580 ± 0.17
		H	0.560 ± 0.021	0.538 ± 0.017	0.555 ± 0.13	0.540 ± 0.12
		Sa	0.539 ± 0.015	0.605 ± 0.030	0.445 ± 0.15	0.455 ± 0.13
		Su	0.555 ± 0.008	0.555 ± 0.018	0.624 ± 0.13	0.494 ± 0.16
	EF (RSVM)	A	0.555 ± 0.021	0.581 ± 0.037	0.320 ± 0.25	0.390 ± 0.18
		D	0.535 ± 0.024	**0.622 ± 0.025**	0.521 ± 0.18	0.500 ± 0.24
		F	**0.618 ± 0.011**	0.619 ± 0.041	**0.700 ± 0.22**	0.522 ± 0.18
		H	0.575 ± 0.017	0.597 ± 0.009	0.575 ± 0.13	**0.524 ± 0.14**
		Sa	0.540 ± 0.021	0.598 ± 0.024	0.532 ± 0.16	0.502 ± 0.14
		Su	0.579 ± 0.013	0.574 ± 0.014	0.433 ± 0.13	0.515 ± 0.16
	LF(RSVM)	A	0.543 ± 0.062	0.571 ± 0.081	0.570 ± 0.15	0.590 ± 0.09
		D	0.542 ± 0.044	**0.597 ± 0.093**	0.590 ± 0.10	0.570 ± 0.14
		F	0.519 ± 0.029	0.597 ± 0.170	**0.645 ± 0.16**	**0.691 ± 0.12**
		H	0.526 ± 0.031	0.508 ±0.017	0.564 ± 0.09	0.552 ± 0.10
		Sa	**0.573 ± 0.076**	0.584 ± 0.102	0.513 ± 0.04	0.580 ± 0.11
		Su	0.562 ± 0.073	0.568 ± 0.125	0.581 ± 0.08	0.583 ± 0.08

Note: Bold numbers indicate best accuracies in predicting the gender of the user using unimodal or multimodal (EEG and gaze) features from given emotion stimuli.

Table 3.3   GR results with CNN and AdaBoost classifiers over different modalities and their fusion

AUC		HI	LI	Eye mask	Mouth mask
ALL	Eye (CNN)	**0.931 ± 0.011**	**0.882 ± 0.020**	0.892 ± 0.038	0.886±0.023
	Eye (Ada)	0.521 ± 0.016	0.538 ± 0.023	0.969 ± 0.015	**0.966 ± 0.014**
	Late Fusion	0.920 ± 0.015	0.850 ± 0.045	**0.974 ± 0.017**	0.946 ± 0.037
EEG (CNN)	A	0.757 ± 0.113	0.719 ± 0.086	0.626 ± 0.165	0.676 ± 0.157
	D	0.758 ± 0.047	0.644 ± 0.080	0.594 ± 0.161	0.673 ± 0.108
	F	0.738 ± 0.083	0.618 ± 0.107	0.542 ± 0.128	0.585 ± 0.199
	H	**0.790 ± 0.049**	0.701 ± 0.054	**0.714 ± 0.082**	**0.751 ± 0.051**
	Sa	0.739 ± 0.089	**0.737 ± 0.078**	0.689 ± 0.097	0.731 ± 0.079
	Su	0.720 ± 0.076	0.636 ± 0.042	0.649 ± 0.098	0.652 ± 0.067
EYE (Ada)	A	**0.651 ± 0.064**	**0.651 ± 0.079**	0.840 ± 0.160	0.957 ± 0.038
	D	0.591 ± 0.038	0.600 ± 0.068	0.950 ± 0.045	0.910 ± 0.108
	F	0.569 ± 0.078	0.528 ± 0.157	0.950 ± 0.067	0.881 ± 0.075
	H	0.574 ± 0.090	0.529 ± 0.050	0.931 ± 0.049	0.929 ± 0.051
	Sa	0.587 ± 0.087	0.576 ± 0.068	**0.959 ± 0.051**	**0.958 ± 0.041**
	Su	0.530 ± 0.093	0.546 ± 0.052	0.938 ± 0.067	0.921 ± 0.107
LF	A	0.720 ± 0.082	0.651 ± 0.113	0.835 ± 0.167	0.860 ± 0.132
	D	0.676 ± 0.083	0.557 ± 0.153	**0.939 ± 0.070**	**0.967 ± 0.105**
	F	0.675 ± 0.140	0.470 ± 0.164	0.887 ± 0.139	0.869 ± 0.097
	H	**0.770 ± 0.050**	**0.683 ± 0.088**	0.897 ± 0.042	0.926 ± 0.060
	Sa	0.729 ± 0.062	0.555 ± 0.107	0.939 ± 0.062	0.843 ± 0.133
	Su	0.723 ± 0.090	0.604 ± 0.105	0.826 ± 0.142	0.943 ± 0.058

Note: Bold numbers indicate best accuracies in predicting the gender of the user using unimodal or multimodal (EEG and gaze) features from given emotion stimuli.

(Emotion-wise) are employed for model training. Also, the best EF and LF results along with the relevant classifier are specified for all conditions. Cumulatively, Tables 3.2 and 3.3 clearly convey that the CNN and AdaBoost frameworks considerably outperform baseline classifiers.

Focusing on Table 3.2, one can note that the AUC scores with HI emotions are typically higher (cf. columns 1 and 2); also, AUC metrics achieved in the mouth-masked condition are typically higher than with eye mask (columns 3 and 4). These findings from implicit user behavior mirror explicit recognition results in Figure 3.4. Specifically, significant differences can be noted with EEG-based results for the above conditions, while differences with eye-gaze are inconspicuous. These results cumulatively convey that (a) analyses of explicit and implicit user behavior affirm similar trends relating to visual FER; (b) gender differences are better encoded while perceiving HI emotions, and how eye cues are interpreted for emotion inference, and (c) EEG signals better encode gender differences in visual emotion processing than gaze patterns.

Examining fusion results, LF is generally superior to EF when all emotions are considered, while EF achieves superior performance when emotion-specific user data are utilized for GR. We attempted GR from emotion-specific data to follow up on findings in Section 1.4, conveying that females are more sensitive to negative emotions. In Table 3.2, eye-gaze-based GR performance improves significantly when emotion-specific data are considered (peak AUC of 0.68 for fear under eye mask vs peak AUC of 0.53 under mouth mask for all emotions), while EEG results remain stable (peak AUC of 0.71 for HI anger vs peak AUC of 0.71 for all HI emotions). Optimal GR results (in bold) are mostly achieved for the A, D, F, and Sa emotions implying that gender differences best manifest for negative valence.

Table 3.3 largely replicates the trends noted in Table 3.2, with higher AUC scores achieved with the CNN and AdaBoost classifiers. The one notable difference is that while the optimal NB baseline performs poorly with gaze features but much better with EEG in Table 3.2, the AdaBoost framework outperforms the EEG-based CNN in the eye and mouth mask conditions; EEG nevertheless encodes gender differences better while processing HI and LI emotions. These results imply that (a) eye movements in pursuit of FER, especially under partial face occlusion, are distinctive of gender, and (b) the discriminative AdaBoost framework can better learn distinctive eye movement patterns as compared to the generative NB classifier. Late fusion results denoting the combination of decisions made by the CNN and AdaBoost reveal that when one modality is considerably more potent than the other (EEG $\gg$ Eye for HI and LI, while Eye > EEG for mask), LF does not necessarily outperform constituent modalities.

3.4.4.3 Spatio-temporal EEG analysis

We also examined spatio-temporal characteristics of the EEG signal to examine if (i) gender differences in visual emotion processing are effectively captured by certain electrodes and are attributable to specific (functional) brain lobes and (ii) any time window(s) were critical for GR over 4 s of visual processing.

(a) **Spatial:** We evaluated the ability of each EEG channel (cf. Figure 3.1) to capture gender-discriminative information by feeding the deep network (Figure 3.8) with single-channel EEG input. Consistent with prior findings [20,71,72], we noted optimal GR with the frontal and occipital brain lobes. Single-channel GR followed a trend similar to Table 3.3, with optimal GR achieved for HI emotions, and the worst GR performance noted with mouth mask data. A symmetricity was noted among the optimal EEG channels, namely, AF3 and AF4, and F3 and F4. These results mirror observations relating to the existence of brain hemispheres [73] for ER (Table 3.4).

Table 3.4   Spatial performance evaluation. Parentheses denote the channel for which the best AUC was achieved.

	HI	LI	Eye mask	Mouth mask
1	0.8266 (AF4)	0.8008 (F3)	0.7160 (F4)	0.7135 (F3)
2	0.8200 (F3)	0.772 (O2)	0.7143 (F3)	0.7124 (F4)
3	0.8131 (O2)	0.7745 (AF4)	0.7055 (AF4)	0.7107 (AF3)

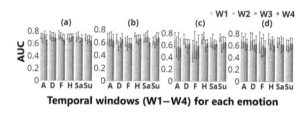

Temporal windows (W1–W4) for each emotion

Figure 3.10   *(a–d) Temporal EEG analyses: GR for HI, LI, eye, and mouth mask conditions over temporal windows (W1–W4). Error bars denote unit standard deviation.*

(b) **Temporal:** As isolation of gender-specific ERPs is possible from 1 s time windows, we considered four non-overlapping 1 s windows W1–W4 spanning 4 s of stimulus viewing. Figure 3.10 presents GR AUCs achieved over W1–W4 from emotion-specific EEG data. Plots confirm that reliable, above-chance GR is achieved over each of W1–W4 across the different viewing conditions. The highest GR performance is noted for HI morphs and the lowest GR for eye mask faces, consistent with Table 3.3. While temporal analyses revealed no significant GR differences across windows, optimal GR was generally achieved for W1 and W2 over all morph conditions suggesting a primacy effect. Masked conditions result in higher GR performance variance, perhaps due to the initial difficulty in FER owing to facial occlusions.

3.5   Discussion and conclusion

A critical requirement of today's ubiquitous computing devices is to sense user intention, emotion, and cognition from multimodal cues, and devise effective interactions to optimize users' individual and social behaviors. Affective computing (AC), whose objective is to empower devices to interact naturally and empathetically with users, plays a crucial role here. Apart from inferring user emotions, it would also be beneficial for AC systems to predict soft biometrics such as user age and gender for effective interaction [74]. There is also a strong need to develop sensing mechanisms that respect user privacy concerns [75], and eye movements and EEG signals represent privacy-preserving implicit behaviors that enable inference of user traits even as the user identity remains hidden.

The primary objective of this study is to explore the utility of eye movements and EEG for user gender prediction. We designed a study where these implicit user behaviors were recorded as 28 users (14 male) recognized facial emotions; since eye movements are known to be characteristic of the perceived facial emotion [25,45], our study design additionally enabled recognition of the perceived facial (stimulus) valence. Crucially, our study employs lightweight and inexpensive sensors for inference, as against bulky and intrusive lab sensors which are typically used in user-centered analyses but significantly constrain user behavior. Also, to examine user efficacy for FER under occlusions, we presented both unoccluded (high and low intensity) and occluded (with the eye or mouth region masked) emotive faces to users.

The fact that gender differences exist in visual emotional processing is demonstrated at multiple levels via our experiments. In terms of explicit user responses, female users are found to perform FER quickly and accurately on unoccluded faces, especially for negative valence emotions (Section 4). Likewise, females also achieve higher recognition of negative emotions with mouth-masked faces. Subsequent examination of implicit cues revealed interesting correlations; examination of eye fixation distributions showed greater female fixation around the eyes than males irrespective of stimulus type, and fixations on eyes were significantly longer in the mouth-masked condition. This observation is consistent with prior studies [23], and the proposition that females primarily look at the eyes for emotional cues.

Analysis of EEG ERPs also conveyed interesting patterns. While processing unoccluded faces, stronger N100 and P300 peaks are noted in female ERPs for negative emotions, as well as a stronger N400 peak for strongly negative faces. This suggests a differential cognitive processing of negative vs positive emotions by females, which results in their enhanced sensitivity toward negative emotions. Likewise, lower male N400 latencies are generally noted at the O2 electrode for positive emotions. Stronger N100 and P300 female ERPs are also noted for negative mouth-masked emotions, suggesting that negative emotions are processed rapidly by women, and do not necessarily entail selective attention to emotional cues. The fact that implicit and differential behaviors can be isolated using data acquired via low-cost sensors affirms that our experimental design can effectively capture emotion and gender-specific information. Greater female sensitivity to negative emotions can be attributed to several factors like social structure, environment and evolution, way of living, and social stereotypes [76]. While we did not seek to expressly elicit emotions through facial imagery, facial expressions are known to induce emotions in the viewer [77], and examination of users' emotional behavior enables prediction of both the user gender and stimulus valence.

VR experiments employing eye-gaze features (processed by an AdaBoost ensemble) and EEG features (input to a three-layer CNN performing 1D convolutions) revealed the following. Eye-gaze patterns on HI emotional faces were highly characteristic of facial valence for both male and female users, resulting in AUC scores ≥ 0.98. Contrastingly, EEG produced largely near-chance performance, implying that the EEG features are sub-optimal for ER. Gulf in the efficacy of eye-gaze and EEG features meant that late fusion of the classifier outputs was hardly beneficial. That human eye movements are highly characteristic of stimulus valence is unsurprising, with prior studies [21] achieving better-than-chance accuracy with gaze features compiled for diverse scenes; on the contrary, our study is specific to emotional faces.

GR results are summarized as follows: Table 3.2 presenting GR results achieved with the baseline NB, LSVM, and RSVM classifiers affirms that consistent with the recognition rate statistics and ERP analyses, EEG-based gender differences best manifest for HI emotional faces and are least observable for LI emotional faces. Emotion-specific analyses affirm that peak GR AUC scores are mostly achieved with negative valence data. Also, the NB classifier achieves much superior results with EEG as compared to eye-gaze features. On the other hand, Table 3.3 shows that much superior GR results are achievable with the CNN and AdaBoost classifiers respectively fed with the EEG and eye-gaze data. Interestingly, very high AUC scores are achieved with eye-gaze features compiled for the mask conditions (this trend is unobservable for the HI and LI conditions), implying that eye movements in pursuit of FER under partial face occlusion are highly gender-specific. The substantial disparity in EEG and eye-based results across the different conditions results in the fusion of the two modalities is hardly beneficial.

While the presented study involves only a small pool of ($N = 28$) users, the observed GR and ER results are nevertheless highly promising and demonstrate the utility of implicit behaviors for privacy-preserving user profiling. The ever-increasing and commonplace availability of sensors employed in this work also opens the possibility of conducting large-scale (crowd-sourced) user-centric studies. Our larger endeavor is to predict soft biometrics (age, gender, emotional, and cognitive state) of users via implicit and multimodal behavioral cues to empower gaming, advertising, augmented and virtual reality applications for behavioral change, and mental health monitoring systems for disorders like Alexithymia. Future work will focus on the modeling of shared behaviors (joint embedding of EEG and eye-gaze features) for efficient user-trait prediction employing techniques such as multi-task learning and prototyping real-life profiling applications. Investigating the optimality of pre-designed features (e.g., power-spectral density for EEG) or learned feature descriptors represents another interesting line of work.

References

[1]   Wiedenbeck S, Grigoreanu V, Beckwith L, and Burnett M. Gender HCI: What about the software? *Computer* 2006;39:97–101.
[2]   Picard RW. *Affective Computing*, Cambridge, MA: MIT Press, 1997.
[3]   Passig D and Levin H. The interaction between gender, age, and multimedia interface design. *Education and Information Technologies* 2001;6(4):241–50.
[4]   Czerwinski M, Tan DS, and Robertson GG. Women take a wider view. In: *Conference on Human Factors in Computing Systems (CHI)*, 2002, pp. 195–202.
[5]   Zhang W, Smith ML, Smith LN, and Farooq AR. Gender and gaze gesture recognition for human-computer interaction. *Computer Vision and Image Understanding* 2016;149:32–50.
[6]   Homer BD, Hayward EO, Frye J, and Plass JL. Gender and player characteristics in video game play of preadolescents. *Computers in Human Behavior* 2012;28(5):1782–89.
[7]   Schwark JD, Dolgov I, Hor D, and Graves W. Gender and personality trait measures impact degree of affect change in a hedonic computing paradigm. *International Journal Human-Computer Interaction* 2013;29(5):327–37.

[8]   Ng CB, Tay YH, and Goi BM. Vision-based human gender recognition: A survey. CoRR. 2012, abs/1204.1611.

[9]   Joho H, Staiano J, Sebe N, and Jose JM. Looking at the viewer: Analysing facial activity to detect personal highlights of multimedia contents. *Multimedia Tools and Applications* 2011;51(2):505–23.

[10]   Li M, Han KJ, and Narayanan S. Automatic speaker age and gender recognition using acoustic and prosodic level information fusion. *Computer Speech & Language* 2013;27(1):151–67.

[11]   Lee CM and Narayanan SS. Toward detecting emotions in spoken dialogs. *IEEE Transactions on Speech and Audio Processing* 2005;13(2):293–303.

[12]   Biometric Security Poses Huge Privacy Risks, 2013. Available at: https://www.scientificamerican.com/article/biometric-security-poses-huge-privacy-risks/.

[13]   Campisi P and La Rocca D. Brain waves for automatic biometric-based user recognition. *IEEE Transactions on Information Forensics and Security* 2014; 9(5):782–800.

[14]   Yang S and Deravi F. On the effectiveness of EEG signals as a source of biometric information. In: *Emerging Security Technologies*, IEEE, 2012, pp. 49–52.

[15]   van Vliet M, Robben A, Chumerin N, Manyakov NV, Combaz A, and Hulle MMV. Designing a brain–computer interface controlled video-game using consumer grade EEG hardware. In: *Biosignals and Biorobotics Conference (BRC)*, 2012, pp. 1–6.

[16]   Lim WL, Sourina O, Wang L. Mind—An EEG neurofeedback multitasking game. In: *Cyberworlds*, IEEE, 2015, pp. 169–72.

[17]   Dalmaijer E. Is the low-cost EyeTribe eye tracker any good for research? PeerJ PrePrints, 2014.

[18]   Montagne B, Kessels RPC, Frigerio E, de Haan EHF, and Perrett DI. Sex differences in the perception of affective facial expressions: Do men really lack emotional sensitivity? *Cognitive Processing* 2005;6(2):136–41.

[19]   Hall JA and Matsumoto D. Gender differences in judgments of multiple emotions from facial expressions. *Emotion* 2004;4(2):201–16.

[20]   Bilalpur M, Kia SM, Chua TS, and Subramanian R. Discovering gender differences in facial emotion recognition via implicit behavioral cues. arXiv preprint arXiv:170808729, 2017.

[21]   Tavakoli RH, Atyabi A, Rantanen A, Laukka SJ, Nefti-Meziani S, and Heikkilä J. Predicting the valence of a scene from observers' eye movements. *PLoS ONE* 2015;10(9):1–19.

[22]   Vassallo S, Cooper SL, and Douglas JM. Visual scanning in the recognition of facial affect: Is there an observer sex difference? *Journal of Vision* 2009;9 (3):11.1–11.10.

[23]   Wells LJ, Gillespie SM, and Rotshtein P. Identification of emotional facial expressions: Effects of expression, intensity, and sex on eye gaze. *PLoS One* 2016;11(12):e0168307.

[24] Subramanian R, Shankar D, Sebe N, and Melcher D. Emotion modulates eye movement patterns and subsequent memory for the gist and details of movie scenes. *Journal of Vision* 2014;14(3):1–18.

[25] Schurgin MW, Nelson J, Iida S, Ohira H, Chiao JY, and Franconeri SL. Eye movements during emotion recognition in faces. *Journal of Vision* 2014;14 (13):1–16.

[26] Bilalpur M, Kia SM, Chawla M, Chua TS, and Subramanian R. Gender and emotion recognition with implicit user signals. In: *International Conference on Multimodal Interaction*, 2017, pp. 379–87.

[27] Zheng WL, Dong BN, and Lu BL. Multimodal emotion recognition using EEG and eye tracking data. In: *Engineering in Medicine and Biology Society Conference (EMBC)*, 2014, pp. 5040–43.

[28] Liu W, Zheng WL, and Lu BL. Multimodal emotion recognition using multimodal deep learning. arXiv preprint arXiv:160208225, 2016.

[29] Abadi M, Subramanian R, Kia S, Avesani P, Patras I, and Sebe N. DECAF: MEG-based multimodal database for decoding affective physiological responses. *IEEE Transactions on Affective Computing* 2015;6(3):209–22.

[30] Koelstra S, Mühl C, Soleymani M, *et al.* DEAP: A database for emotion analysis using physiological signals. *IEEE Transactions on Affective Computing* 2012;3(1):18–31.

[31] Subramanian R, Wache J, Abadi M, Vieriu R, Winkler S, and Sebe N. ASCERTAIN: Emotion and personality recognition using commercial sensors. *IEEE Transactions on Affective Computing* 2018;9(2):147–60.

[32] Wu Y, Zhuang Y, Long X, Lin F, and Xu W. Human gender classification: A review. arXiv preprint arXiv:150705122. 2015.

[33] Bilalpur M, Kankanhalli M, Winkler S, and Subramanian R. EEG-based evaluation of cognitive workload induced by acoustic parameters for data sonification. In: *International Conference on Multimodal Interaction*, 2018, pp. 315–23.

[34] Hoppe S, Loetscher T, Morey SA, and Bulling A. Eye movements during everyday behavior predict personality traits. *Frontiers in Human Neuroscience* 2018;12(105):328195.

[35] Zotto MD and Pegna AJ. Processing of masked and unmasked emotional faces under different attentional conditions: An electrophysiological investigation. *Frontiers in Psychology* 2015;6:1691.

[36] Katti H, Subramanian R, Kankanhalli M, Sebe N, Chua TS, and Ramakrishnan KR. Making computers look the way we look: exploiting visual attention for image understanding. In: *ACM Multimedia*, 2010, pp. 667–70.

[37] Hanjalic A and Xu LQ. Affective video content representation and modeling. *IEEE Transactions on Multimedia* 2005;7(1):143–54.

[38] Wang HL and Cheong LF. Affective understanding in film. *IEEE Transactions on Circuits and Systems for Video Technology* 2006;16(6):689–704.

[39] Vonikakis V, Subramanian R, Arnfred J, and Winkler S. A probabilistic approach to people-centric photo selection and sequencing. *IEEE Transactions on Multimedia* 2017;19(11):2609–2624.

[40] Shukla A, Gullapuram SS, Katti H, Yadati K, Kankanhalli M, and Subramanian R. Affect recognition in ads with application to computational advertising. arXiv preprint arXiv:170901683, 2017.

[41] Li X, Zhang P, Song D, Yu G, Hou Y, and Hu B. EEG based emotion identification using unsupervised deep feature learning. In: *Workshop on Neuro-Physics Methods in IR*, 2015.

[42] Zheng WL, Zhu JY, Peng Y, and Lu BL. EEG-based emotion classification using deep belief networks. In: *International Conference on Multimedia and Expo (ICME)*, 2014, pp. 1–6.

[43] Walker-Smith GJ, Gale AG, and Findlay JM. Eye movement strategies involved in face perception. *Perception* 1977;6:313–26.

[44] Smith ML, Cottrell GW, Gosselin F, and Schyns PG. Transmitting and decoding facial expressions. *Psychological Science* 2005;16(3):184–89.

[45] Subramanian R, Yanulevskaya V, and Sebe N. Can computers learn from humans to see better? Inferring scene semantics from viewers' eye movements. In: *ACM Multimedia*, 2011, pp. 33–42.

[46] Aviezer H, Hassin RR, Ryan J, *et al.* Angry, disgusted, or afraid? *Psychological Science* 2008;19(7):724–32.

[47] Etcoff NL and Magee JJ. Categorical perception of facial expressions. *Cognition* 1992;44(3):227–40.

[48] Sullivan S, Campbell A, Hutton SB, and Ruffman T. What's good for the goose is not good for the gander: Age and gender differences in scanning emotion faces. *Journals of Gerontology, Series B: Psychological Sciences and Social Sciences* 2015:1–6.

[49] Bassili JN. Emotion recognition: The role of facial movement and the relative importance of upper and lower areas of the face. *Journal of Personality and Social Psychology* 1979;37(11):2049–58.

[50] Hall JK, Hutton SB, and Morgan MJ. Sex differences in scanning faces: Does attention to the eyes explain female superiority in facial expression recognition? *Cognition and Emotion* 2010;24(4):629–37.

[51] Rahman Q, Wilson GD, and Abrahams S. Sex, sexual orientation, and identification of positive and negative facial affect. *Brain and Cognition* 2004;54(3):179–85.

[52] Lithari C, Frantzidis CA, Papadelis C, *et al.* Are females more responsive to emotional stimuli? A neurophysiological study across arousal and valence dimensions. *Brain Topography* 2010;23(1):27–40.

[53] Nguyen P, Tran D, Huang X, and Ma W. Age and gender classification using EEG paralinguistic features. In: *Conference on Neural Engineering (NER)*, IEEE, 2013. pp. 1295–98.

[54] Turkstra LS, Mutlu B, Ryan CW, *et al.* Sex and gender differences in emotion recognition and theory of mind after TBI: A narrative review and directions for future research. *Frontiers in Neurology* 2020;11:59.

[55] Saylik R, Raman E, and Szameitat AJ. Sex differences in emotion recognition and working memory tasks. *Frontiers in Psychology* 2018;9:1072.

[56] Fitriani M, Khotimah S, Haryanto F, and Suprijadi S. Study of electroencephalogram pattern from eye response to flickering light. *Journal of Physics: Conference Series* 2019;4:1204.

[57] Langner O, Dotsch R, Bijlstra G, Wigboldus DHJ, Hawk ST, and van Knippenberg A. Presentation and validation of the Radboud Faces Database. *Cognition and Emotion* 2010;24(8):1377–88.

[58] Baltru˘saitis T, Robinson P, and Morency LP. OpenFace: An open source facial behavior analysis toolkit. In: *Winter Conference on Applications of Computer Vision (WACV)*, 2016.

[59] Muhl C, Allison B, Nijholt A, and Chanel G. A survey of affective brain computer interfaces: principles, state-of-the-art, and challenges. *Brain-Computer Interfaces* 2014;1(2):66–84.

[60] Vi CT, Jamil I, Coyle D, and Subramanian S. Error related negativity in observing interactive tasks. In: *SIGCHI Conference on Human Factors in Computing Systems*, 2014. pp. 3787–96.

[61] Krassanakis V, Filippakopoulou V, and Nakos B. EyeMMV toolbox: An eye movement post-analysis tool based on a two-step spatial dispersion threshold for fixation identification. *Journal of Eye Movement Research* 2014;7(1).

[62] Koelstra S and Patras I. Fusion of facial expressions and EEG for implicit affective tagging. *Image and Vision Computing* 2013;31(2):164–74.

[63] Jeni LA, Cohn JF, and De La Torre F. Facing imbalanced data–recommendations for the use of performance metrics. In: *2013 Humaine Association Conference on Affective Computing and Intelligent Interaction*, IEEE, 2013, pp. 245–51.

[64] Rad NM, Kia SM, Zarbo C, *et al.* Deep learning for automatic stereotypical motor movement detection using wearable sensors in autism spectrum disorders. *Signal Processing* 2018;144:180–91. Available at: http://www.sciencedirect.com/science/article/pii/S0165168417303705.

[65] Ioffe S and Szegedy C. Batch normalization: Accelerating deep network training by reducing internal covariate shift. In: Bach F and Blei D (editors). *International Conference on Machine Learning*, vol. 37, 2015, pp. 448–56. Available at: http://proceedings.mlr.press/v37/ioffe15.html.

[66] Simonyan K and Zisserman A. Very deep convolutional networks for large-scale image recognition. CoRR, 2014, abs/1409.1556.

[67] Chollet F, *et al.* Keras. GitHub, 2015. https://github.com/fchollet/keras.

[68] Viola P and Jones MJ. Robust real-time face detection. *International Journal on Computer Vision* 2004 May;57(2):137–54. Available at: https://doi.org/10.1023/B:VISI.0000013087.49260.fb.

[69] Silapachote P, Karuppiah DR, and Hanson AR. Feature selection using AdaBoost for face expression recognition. *UMASS Department of Computer Science*, 2005.

[70] Hastie T, Rosset S, Zhu J, and Zou H. Multi-class AdaBoost. *Statistics and Its Interface* 2009;2(3):349–60.

[71] Bilalpur M, Kia SM, Chawla M, Chua TS, and Subramanian R. Gender and emotion recognition with implicit user signals. In: *International Conference on Machine Learning*, 2017.

[72] McClure EB, Monk CS, Nelson EE, *et al.* A developmental examination of gender differences in brain engagement during evaluation of threat. *Biological Psychiatry* 2004;55:1047–55.

[73] Vingerhoets G, Berckmoes C, and Stroobant N. Cerebral hemodynamics during discrimination of prosodic and semantic emotion in speech studied by transcranial Doppler ultrasonography. *Neuropsychology* 2003;17(1):93.

[74] Rukavina S, Gruss S, Hoffmann H, Tan JW, Walter S, and Traue HC. Affective computing and the impact of gender and age. *PloS One* 2016; 11(3):1–20.

[75] Reynolds C and Picard R. Affective sensors, privacy, and ethical contracts. In: *CHI'04 Extended Abstracts on Human Factors in Computing Systems*, 2004, pp. 1103–06.

[76] Deng Y, Chang L, Yang M, Huo M, and Zhou R. Gender differences in emotional response: Inconsistency between experience and expressivity. *PLoS One* 2016;11(6):e0158666.

[77] Siedlecka E and Denson TF. Experimental methods for inducing basic emotions: A qualitative review. *Emotion Review* 2019;11(1):87–97.

Chapter 4

Gesture-oriented supernumerary robotic fingers for post-stroke rehabilitation

Chetana Krishnan[1], Vijay Jeyakumar[1], A. Anusha[1] and S. Shivapriya[1]

The National Health Portal claims that around 70% of the world's rehabilitation population are amputees with minimum residual capacity who cannot fully use the therapies. The existing technology provides implants and prosthetic devices for those with negligible residual capacity, but the same does not work for semi-amputees. The proposed method tends to develop affective computing-based robotic fingers to help patients in their day-to-day activities. The novel algorithm is an extended version of Hexosys base with more complex variables and coefficients in action to improve the gloves' accuracy and flexibility. In the first phase of the system, laptop webcams capture the required gesture patterns, and the computing algorithm processes them to get a numerical pattern. In the second phase, the glove offers customized finger flexion and extension with six degrees of freedom by a set of encoders and actuators attached to the control system. The system can be further extended to pattern complex multiple gestures to increase the effectiveness of the therapy. This chapter will demonstrate a preliminary study to understand the effectiveness of the affective computing algorithm integrated with Hexosys to provide body-powered-assistive training to people. The primary training was done in the healthy control groups and an accuracy of 93.45 was achieved. The proposed model replicated the baseline model with a 0.89 correlation score.

Keywords: Assistive technology; Human–machine interaction; Gesture; Stroke rehabilitation

4.1 Introduction

A medical condition characterized by the interruption of blood supply to the brain caused due to blocks or other factors. It leads to impairment in walking, speaking,

[1]Department of Biomedical Engineering, Sri Sivasubramaniya Nadar College of Engineering, India

and understanding. Later effects include stroke integrated with paralysis and amputation due to numbness, and the latter does not lead to complete paralysis. Still, a minimal residual capacity is functional in both the targeted and the healthy hand, thus opening opportunities for brain–robotic interface systems. The common types of strokes are:

- **Ischemic stroke** is where severe clots block the blood supply to the brain, causing numbness in the origination and leading to semi-amputation. About 43%–56% of residual capacity (the ability to perform physiological detrital activities) is functional in this case.
- **A transient stroke** occurs when the blocked region ruptures due to heavy pressure and hand loading, leading to full automation. A negligible amount of residual capacity is functional in this case.

4.1.1   Post-stroke rehabilitation

Post-stroke rehabilitation typically starts as soon as the stroke is detected and established. This is a crucial step as it prevents repetitive occurrences of stroke. Also, this dramatically reduces the possibility of paralysis and generalizes mobilization along with doing routine chores. Rehabilitation is achieved through various methods and focuses on upper limb rehabilitation, gait rehabilitation, cognitive rehabilitation, and speech and language rehabilitation. Yoga and spiritual training have become an integral part of rehabilitation. The use of prosthetics and assistive devices has increased multifold in recent years. Domains such as stem cell therapy and herbal treatments are being experimented with for post-stroke rehabilitation [1].

4.1.2   Methods existing for post-stroke rehabilitation

The rehabilitation of stroke is an integrated technique that includes people from varying domains in healthcare. The objective is to decrease the number of survivors with neurological and functional impairments. Hydrotherapy means immersion in water, a therapeutic therapy that is highly efficient physically and mentally. Performing exercises in warm water helps in the improvement of musculoskeletal conditions. The temperature and other properties must be considered before the therapy is taken.

Physical therapy has been playing a vital role for ages. Yet, more of a holistic module is evolving, including motor and spasticity management, sensory and cognitive impairment, nutrition, and depression. Botulinum Toxin-A (BoNT-A) has been widely used to treat spasticity in recent times. The most common solution to the above problem is using prosthetic devices that work either by body-powered or muscle-powered mechanisms. The limitations like non-reliability, cost, heaviness in material, and inaccuracy to perfectly mimic human physiology are vast causes of less acceptance [2].

4.1.3   Modern techniques for rehabilitation

At present, robotics is playing an essential part in neurorehabilitation. End effector (EE) type and exoskeleton variants are the two major classifications of robots used. A study suggests that treatment through the EE robot has proved to be much more helpful in the cases of chronic stroke than the exoskeleton method. Robotic-assisted devices with virtual and augmented reality have also come into existence. They have proved to significantly assist rehabilitation as they deliver high-intensity training and prove adaptive. The characteristic feature of such assistive devices is that they produce simulations that provide visual and multi-sensory feedback. These methods lack cost effectiveness which in turn reduces their usage.

Electrical stimulation (ES) is a growing therapy approach and has proven effective. Neuromuscular ES supplies pulse stimulation of higher energy over the required muscle region with the aid of surface electrodes. The skeletal muscles have recovered at a greater rate with this technique. The transcutaneous electrical neuromuscular stimulation (TENS) unit has been designed to use low-voltage electrical signals through surface electrodes. Two theories have been proposed to explain the working of TENS. The first is blocking pain signals' transmission by giving electrical signals and stimulating nerve cells. This excitation of neurons leads to the release of endorphins that change the perception of pain. TENS is low-cost and user-friendly [3].

4.2   Affective computing in healthcare

Affective computing for rehabilitation gains specific attention in the growing technology world as a broad spectrum of development is seldom found without being integrated with emotions [4]. This field mainly focuses on enhancing the user's experience with the concerned machine with the help of artificial intelligence (AI) and the Internet of Things (IoT). Affective computing ensures that the feelings received are quantified accordingly to ensure that devices can perceive them. Recent breakthroughs in the same field include emotion recognition and sentiment analysis. Physical emotions being expressed may not always be true to one's conscience. Hence, there is a fusion of physical information and physiological signals to obtain valid results.

4.2.1   Affective computing for post-stroke rehabilitation

As discussed earlier, affective computing plays a significant role in interpreting our emotions; hence they act as great treatment options for interaction and communication-related disorders. Among the effective remedies for post-traumatic stress disorder, affective computing is frequently sought. This concept is illustrated in Figure 4.1.

Electroencephalogram (EEG) is a nonlinear analysis that gives us better insight into the emotional status of stroke patients. Thus, higher-order spectra are being utilized for finding the affective states. The emotional states of stroke patients are determined with the help of bispectral analysis which is nothing but a third-order spectrum [5].

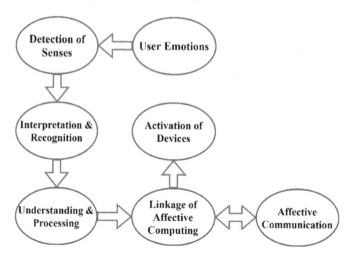

Figure 4.1 Affective computing in healthcare

4.3 Finger anatomical dynamics

Dynamics play a pivotal role in the function of our hands. The hand consists of many articulate surface joints. The coordination and regulation of these joints, along with the muscles and ligaments, allows us to manipulate our hand's movements and perform the required functions. The sensory feedback mechanism, including the brain, muscles, and tendons of the fingers, helps in the controlled activity of the hand. The primary anatomical configuration of the hand comprises eight carpal bones which articulate with five metacarpals and three phalanges in all the fingers except for two in each thumb. These bones are attached through joints which may be classified into three groups:

Carpometacarpal (CMC), metacarpophalangeal (MCP), and interphalangeal (IP) joints. These facilitate finger movements such as flexion, extension, and adduction, which occur around the transverse and anteroposterior axes. These axes are considered to be parallel to each other [6]. The CMC joint of the thumb is a saddle-type joint; thus, a more excellent range of movement can be expected in the thumb, and little finger, whereas the ligament attached to this joint in the other fingers does not encourage much activity [7,8].

Grasps can be of various types like power and precision grasps. The combination of multiple grasps and pinches makes performing activities of daily living (ADLs) possible. The skin of the hand is unique, and the dorsal skin is loosely packed while the palmer skin is thicker, and the reason behind this is to handle more pressure from objects and hold tight on them. The pulley system has flexor tendons against the phalanges, thus preventing them from breaking down [8–10].

4.4   How is affective computing better than the current models?

Affective computing has its edge over other existing models. Using computational systems makes it possible to coordinate a diverse range of data. Affective computational models are known to make dynamic interactions more concrete. They also assist in an improved conceptualization of rehabilitation studies and a better understanding of its design studies, thus enhancing speedy recovery. The usage of affective computing also aids in optimizing the therapy in context with quantity, timing, and location, which signifies the precision of this model. They also provide a better prognosis by observing the various parameters and other gestures involved. By using affective computing, means of computer technology can infiltrate human society comprehensively. It is also possible to provide enhanced human–computer interaction. We can guarantee the safety of equipment or a machine in affective computing (for instance, when a driver is not in a proper mental state to drive a vehicle, it can be detected accordingly). Smart appliances are known to function better when they are aware of the affective state of their user.

4.5   Current prosthetic models and their limitations

Constraint-induced movement therapy (CIMT) is a well-acclaimed therapy for improving forearm motor functioning after a stroke. The idea behind this is to use the affected limb rather than the unaffected limb so that the affected limb starts to perform better with time. The victims generally force the non-paretic limb with the help of a glove or sling for a stroke. There is an age constraint for the usage of CIMT. This can be efficient for people with only one affected arm and having a set degree of activity, on the other hand [8].

Cortimo system has been developed that comprises a brain implant that measures the brain's activity using a sensor and a wearable device for the forearm. It is a battery-powered arm orthosis. It consists of two percutaneous multiports, multi-electrode arrays, amplifiers, and sensors. The sensors are made up of silicon microelectrodes [9]. They penetrate the brain's cortex, measure neuron activity, send electrical signals as required, and improve neurorehabilitation. They are cost-ineffective.

4.6   Affective computing in gesture therapy

HCI contains a subset known as affective computing which investigates the probability of a computer understanding human emotions. Emotions play a crucial part, as the vital factor in any rehabilitation is to keep oneself motivated, perform the intense training program, and perform ADLs and other routine activities [10].

It is possible to achieve post-stroke rehabilitation with the help of effective computing based on the principles of IoT and AI [10]. Studies have shown that patients undergoing post-stroke rehabilitation spend approximately 47 min in occupational therapy daily, but only 4–11 min were dedicated to upper extremity rehabilitation. While implementing affective computing, it becomes crucial to detect the patient's affective states. Three different methods can detect these affective states:

- Speech
- Body gestures
- Physiological changes inside our body

The motion of the hands, such as flexion, extension, abduction, and adduction, can also be a study region for detecting affective states [11]. Detecting affective states mainly constitutes recognizing various emotions and establishing a model to measure human emotions. This model can be designed using reinforcement learning, Markov process, and deep learning techniques [12]. Initially, they used to proceed by calling in the concerned patients and making them undergo a virtual rehabilitation session known as gesture therapy. The gesture therapy consolidates five interactive modules, namely:

- The hardware system consists of a computer, a webcam, the constructed device, and an object in contact with the patient (say gripper).
- Tracking system
- Simulated environment to control the interactions.
- Trunk compensation detector to observe the patient's movements.
- Adaptation system to monitor the patient's growth and progress and implement real-time adaptations depending on the patient's affective state [13].

Gesture therapy can be advanced by measuring the pressure or fatigue in the affected limb, thus making it a parameter for measuring the affective state. Affective computing can later be integrated into this specific model. This session involves virtual reality games as therapeutic exercises [14]. The modules mentioned above work together to track information about the movements and pressure involved by detecting the 3D coordinates for the hand movement and the finger pressure values, which are fed into the simulated environment. Later this information is used to make real-time amendments to the game [15].

The selection of affective states varies from one person to another. Tiredness, fatigue, and pain are a few common factors recommended.

The models for post-stroke rehabilitation provide a quantitative description of sensorimotor activity achieved during the therapy sessions as input, available in the simulated environment or the wearables. In addition, the output thus obtained from such a model is quantitative, time-variant, and related to the functional outcome. These models are designed in such a way that they are dynamic and also possess functional or biological relevance.

Another essential domain that facilitates post-stroke rehabilitation is the brain–computer interface (BCI). BCIs can provide sensory feedback of current brain oscillations [16], so stroke survivors can modulate their sensorimotor rhythms purposefully. Prolonged use of BCIs induces neurological recovery and hence supplements and improves neurological recovery. So, this is how a BCI works. The electric or magnetic activity of the brain is translated into control signals by BCI. These signals are then fed into an external device that modulates the neutral output, accordingly, thus changing the interaction between the brain and its external and internal environment. BCIs can be used as assist devices, thus helping to restore lost functions and aiding in rehabilitation. The better outcomes of the BCIs are due to their ability to induce neuroplasticity [17].

The objectives of the proposed method are as follows:

- The major part of this system includes the development of a finger dynamic model that takes in, interprets, and processes the gestures (preliminary data) as the raw data. Maximum accuracy has to be ensured to make the model desirable.
- The secondary requirements are integrating affective computing algorithms and AI programming [18] to extract high-level features from the process information and to build a human–computer interface environment.

4.7   System architecture

The design architecture of the workflow in the proposed system is shown in Figure 4.2. The multimodal glove follows this with physiological mimicking features enclosed with electromyography (EMG) and ENG sensors (secondary data) that serve as feedback input to the environment [19]. Accuracy is verified by comparing the simulated model data with real-time data using a K-means algorithm and regression methods.

4.7.1   Extended Hexys model construction

Human fingers constitute around 14 bones in translational motion perspective and have 21 degrees of freedom (DoF) (excluding three used for abduction). The main limitation of the current prosthetic model is to mimic and satisfy all the DoF. The primary objective of the proposed system is to build an anatomical model close to 20 DoF. Figure 4.2 shows the finger linkage system of the proposed model [20]. It shows three significant linkages—L_1, L_2, and L_3 which connect the main joints, proximal interphalangeal joints (PIP), and MCP joints. The sub-linkages connect the main joints with the distal IP joint. The angle of motion and connectivity is represented by θ_1, θ_2, and θ_3.

Their respective labels denote the angle of joint motion and reducibility. The model's output feedback is claimed to originate from the third linkage behind the fingertip (common finger movement). The force vector maintains an angle of 54° with the L_3 vector, thus increasing the PIP joint freedom. "Finger chain" in Figure 4.3 represents the joint axial lattice, and Hexosys is the combination of Hexys base with the extended variable parameters.

The kinematics of the linkage system is governed by the following equation:

$$\theta_1 + L_1' + \theta_{\text{mcp}} = F_1 + O_{\text{E}} \tag{4.1}$$

The physiological angle motion is usually clockwise, but the model was developed to make it anti-clockwise to improve chain rotation and angulation. The static

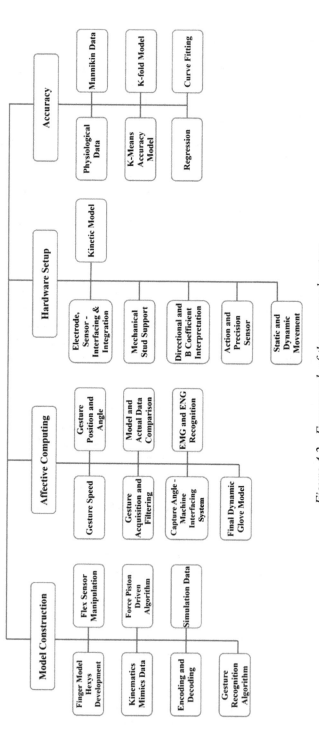

Figure 4.2   Framework of the proposed system

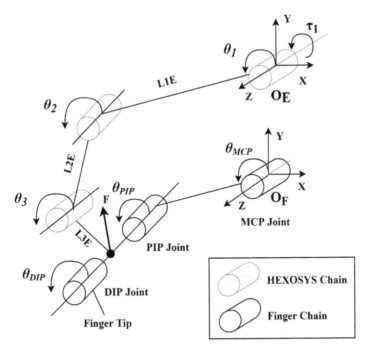

Figure 4.3   Anatomical linkage system

force (F_1) is applied at the hind position of the PIP joint, and the dynamic or the moment force, torque (T_1), is used near the effector muscle region (O_E). Both these forces are parallel to each other, giving the net momentum to lift an object of specified force. The above variables were considered in a single function called L_1 in (4.1). The finger dynamic model proposed is described in Figure 4.3. Once the processor is initialized, the forward movement coefficients like finger abduction and adduction are captured and mapped in the algorithm frame. The corresponding linkage dimensions are fixed. This is followed by acquiring supination, rotation, and pronation movements. The worst-case collision is defined as the minimum deviation of the dynamic model from the actual physiological human model, which plays an essential role in determining the model's accuracy. If this factor is less than zero, the algorithm continues manipulating the impact factor (grasp functions) and dexterity (holding functions). If the collision distance is above zero, the model is calibrated again for variables, and the process is repeated. The additional variable acquired is the finger angle which trains the processor for the appropriate gesture matching.

The finger model was applied in various locations, as shown in Figure 4.4. These locations are fixed by trailing 20 healthy individuals in the age group 30–40. The results were extrapolated to 10 patients of slightly higher age groups who were amputated. The finger rest distance and the corresponding interlocking distance were measured, a part of the encoder and piston setup (Figure 4.5). The green shades represent the most fixational location which tends to remain the same in all

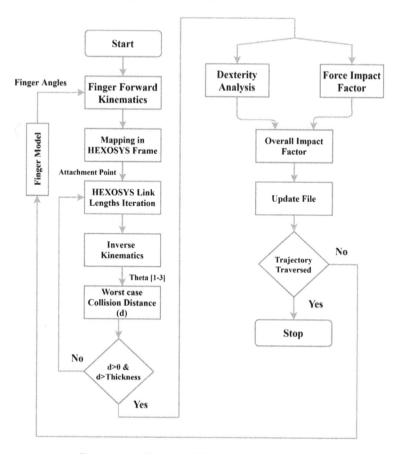

Figure 4.4   Proposed finger dynamics model

the age groups. The red shades represent the changes in age groups due to stiffness and length. The blue shades represent the minority gesture recognitions that help differentiate different gesture patterns. Processing IDE was used to locate, map, and scan the interlocking points, serving as the model's calibration criteria. Most of the impact force is calculated from the tips of the blue region (peak inertial point).

The model can be applied to the lower extremity as well. The transverse trajectory function helps get the biofeedback signals from the EMG electrodes placed under the glove.

The theta value in the model represents the angular motion of the Exo finger, which determines the worst-case collision upon selective criteria [21].

4.7.2   Exoskeleton glove development

The term supernumerary refers to the upper modality features of a physiological system which helps maximize the mimics between the Exo and the anatomical

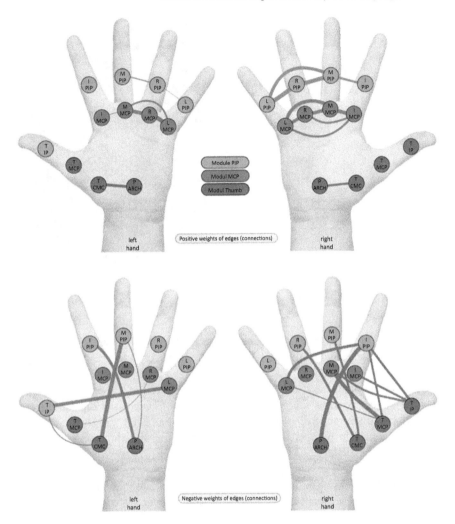

Figure 4.5   Distance manipulation

system [22]. The glove consists of the following functionalities, as shown in Figure 4.6.

- Complex force pattern matrix sensors help balance the force inertia between the encoders. Equation (4.2) governs glove inertia.

$$I = \frac{L}{W} + \frac{I_1}{L_2} \qquad (4.2)$$

where L and L_2 are linkage dimensions, W is the weight of the glove with the arm, and I_1 is the instantaneous inertia at any time [23].

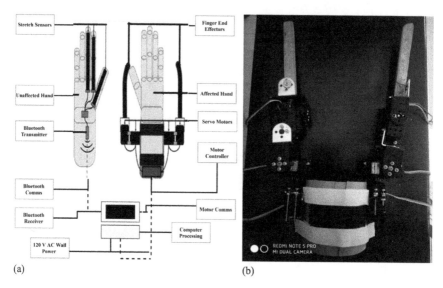

Figure 4.6   (a) Components associated with Exo glove; (b) prototype

- Gesture flex sensors: These sensors are preprogrammed with the proposed finger model, which calculates the isotropy level for every gesture angle produced by the glove.
- The spring damper system is arranged according to the Kelvin Voight model. The system controls the grasp functions of the finger based on the object in action and the range of motion of the Exo fingers [22].
- Encoders: Help process the information from the affecting computing algorithm to the electrodes.
- BOLT module: The interface where the entire algorithm uses a support vector machine model to capture and process the appropriate motion sequence, feed, and rotational vectors to pass it to the spring damper system.

The servo motors help capture and move the Exo based on the EMG recording.

4.7.3   Working algorithm

The EMG electrodes acquire the muscle action signal above 20 millivolts and pass it to the processor for amplification and processing [24]. The features are extracted. The flex sensors acquire the gestures matched with the features to produce an event of movement from the rotary sticks placed in front of the glove. The Bluetooth and the glove interface help the features pass to the amputated side, assisting the patient in producing movements [25]. Figure 4.7 shows the affective training algorithm that the glove incorporates. For the training model, random numbers are subjected to the computer, which the patient is expected to mimic in a timeframe using visual and basic finger movements. Each vibration intensity prompts the level of accuracy, and when resonance is found, the number is changed. This way, the

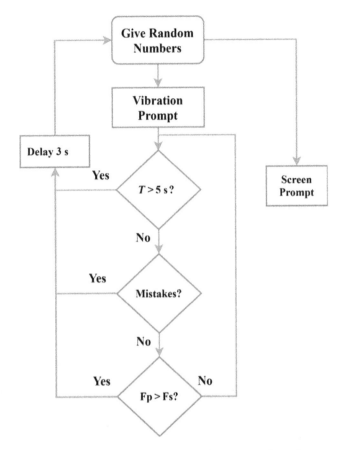

Figure 4.7   Affecting computing training algorithm

complexity increases until the paradigm is trained. Every prompt is followed by a delay of 3 s to allow the extraction. F_p refers to the PIP joint force, and F_s refers to the inter-joint force. Only when $F_p > F_s$ does the model proceed or revert to the calibration phase.

4.8   Results and discussion

The glove was tested on 20 healthy patients (10 testing and 10 training) with good physical ability, non-medicated and gait-dynamized abilities, and 10 amputee patients, and the results were divided into two sections.

4.8.1   Offsite analysis

The main aim of this analysis is to predict the impact: force factor between the subjects, and it was found that the ratio was 2:1 for healthy amputees. This helped

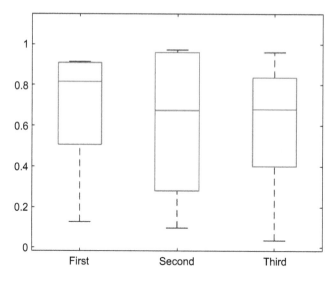

Figure 4.8   Predictive response (X-axis: patient set, Y-axis: level scale)

acquire the EMG at crucial points like the tendon break points to achieve more mimicry. For the selective sensors considered, 40 classes of hand gestures were considered (supination and pronation, to name a few), each falling into different complexity levels. Figure 4.8 shows the predictive response obtained from the three subject sets.

The 40-class EMG classifier performed much better than any other encoder set for all three criteria (the median force was 82.8% for participants who were able-bodied and 79.8% for amputee subjects).

4.8.2   Onsite analysis

The gesture pattern of the three categories was visualized in this case, as seen in Figure 4.8. Comparable to our offsite analysis, the healthy group's highest average dexterity coefficient (DC) was achieved with the sEMG classifier [22]. When only sEMG data was used in such a situation, average DC was substantially higher than for class I ($p < 10^{-2}$) Q test followed by set-wise tests by multiple comparisons. Nine out of the 10 healthy subjects showed the same trend as the one that was observed. Classes I and II showed no discernible differences, even though the average DC for classes II and III was 13%–14% higher than for class I. A slightly different pattern was envisioned for the amputee class. A subset of sensors' sEMG data was employed to produce the best encoding performance in DC. While the internal sensors recorded the activity of the effector muscle group, the outer ones focused on the tendon PIP muscle.

Out of 14 DoF, 10 were achieved by the proposed Exo model, and future work will be to increase the servo collision to increase the DoF factor. The

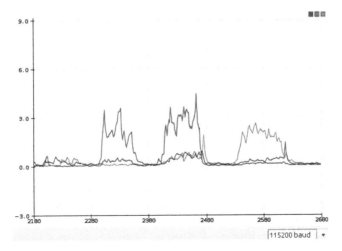

Figure 4.9   EMG classifier example (X-axis: mV and Y-axis: instantaneous time)
(Blue: Class 1 Subjects, Green: Class II Subjects, Red: Class III
Subjects)

patients could promote hand movements with an effective grasping capacity of
about 2.3 Newtons.

4.9   Conclusion

This chapter aims to familiarize the readers with stroke, its consequences, and the
possible rehabilitation available. The role of affective computing and its application
discussed in this chapter positively impact the healthcare domain. The challenges
behind the conventional methods were observed, and the gesture-oriented robotic
fingers will be an effective model for future purposes. The proposed model elapses
exponentially, proving the hypothesis that long-term usage of the Exo glove can help
the patients regain their actual dexterity capacity. To understand the Exo feedback,
12% actuation was established underground force. It was concluded that the healthy
subject (regardless of males and females) could provide the threshold level grasp
force. Still, the test subject's force followed an exponential pattern showing that the
residual capacity dies off due to muscle fatigue. To improve this condition, a central
matrix system is used where the piston takes up the more significant acquisition force
and leaves minimal torque power to the user, thus providing force shifting ability.
The proposed model uses rubber material, making it light in weight and portable.

Authors' contributions

C.K. formulated and devised the proposed Exo model. V.J. validated and tested the
results. A.A and S.S. drafted this manuscript.

Conflicts of interest

The authors have no conflicts of interest.

References

[1] 4P. Sanchetee, Current trends in stroke rehabilitation, *Ischemic Stroke*. London: IntechOpen, 2021. Available:https://www.intechopen.com/chapters/74822 DOI:10.5772/intechopen.95576.

[2] G. N. Yannakakis, Enhancing health care via affective computing, *Malta Journal of Health Sciences*, vol. 5, no. 1, pp. 38–42, 2018.

[3] J. J. Rivas, M. C. Lara, L. Castrejón, *et al.*, Multi-label and multimodal classifier for affective states recognition in virtual rehabilitation, *IEEE Transactions on Affective Computing*, vol. 13, 3, pp. 1183–1194, 2022, doi:10.1109/TAFFC.2021.305579.

[4] Y. Wang, W. Song, W. Tao, *et al.* A systematic review on affective computing: Emotion models, databases, and recent advances, 2022, doi:10.48550/ARXIV.2203.06935.

[5] C. W. Yean, W. K. Wan Ahmad, W. A. Mustafa, *et al.* An emotion assessment of stroke patients by using bispectrum features of EEG signals. *Brain Science* 2020;10(10):672. doi:10.3390/brainsci10100672. PMID: 32992930; PMCID: PMC7601112.

[6] N. Takahashi, S. Furuya, and H. Koike, Soft exoskeleton glove with human anatomical architecture: Production of dexterous finger movements and skillful piano performance, *IEEE Transactions on Haptics*, vol. 13, no. 4, pp. 679–690, 2020, doi:10.1109/TOH.2020.2993445.

[7] H. Mnyusiwalla, P. Vulliez, J. -P. Gazeau, and S. Zeghloul, A new dexterous hand based on bio-inspired finger design for inside-hand manipulation, *IEEE Transactions on Systems, Man, and Cybernetics: Systems*, vol. 46, no. 6, pp. 809–817, 2016, doi:10.1109/TSMC.2015.2468678.

[8] D. J. Reinkensmeyer, E. Burdet, M. Casadio, *et al.*, Computational neurorehabilitation: modeling plasticity and learning to predict recovery, *J Neuro Engineering Rehabilitation*, vol. 13, no. 42, 2016. https://doi.org/10.1186/s12984-016-0148-3.

[9] A. Gustus, G. Stillfried, J. Visser, *et al.*, Human hand modelling: kinematics, dynamics, applications, *Biological Cybernetics*, vol. 106, pp. 741–755, 2012. https://doi.org/10.1007/s00422-012-0532-4.

[10] H. Liu, H. Lieberman, and T. Selker, A model of textual affect sensing using real-world knowledge, *Proceedings of the 8th International Conference on Intelligent User Interfaces*. ACM Press, Miami, FL, pp. 125–132, 2003.

[11] A. Ryan, J. F. Cohn, S. Lucey, *et al.*, Automated facial expression recognition system, *Proceedings 43rd Annual 2009 International Carnahan Conference on Security Technology*. IEEE, Zürich, Switzerland, pp. 172–177, 2009.

[12] F. Yang, C. Li, R. Palmberg, E. Van Der Heide, and C. Peters, Expressive virtual characters for social demonstration games, *2017 9th International Conference on Virtual Worlds and Games for Serious Applications (VS-Games)*, 2017, pp. 217–224, doi: 10.1109/VS-GAMES.2017.8056604.

[13] Kumar, S, Deep learning based affective computing, *Journal of Enterprise Information Management*, vol. 34, no. 5, pp. 1551–1575, 2021, doi: https://doi.org/10.1108/JEIM-12-2020-0536.

[14] S. Bhattacharyya, C. Cinel, L. Citi, D. Valeriani, and R. Poli, Walking improves the performance of a brain–computer interface for group decision making, *Current Research in Neuroadaptive Technology*, pp. 221–233, 2022, DOI:10.1016/B978-0-12-821413-8.00017-8.

[15] L. Chen, Y. Chen, W. B. Fu, D. F. Huang, and W. L. A. Lo, The effect of virtual reality on motor anticipation and hand function in patients with sub-acute stroke: A randomized trial on movement-related potential, *Neural Plasticity*, vol. 2022, pp. 1–14, 2022, https://doi.org/10.1155/2022/7399995.

[16] B. Yang, J. Ma, W. Qiu, Y. Zhu, X. Meng, A new 2-class unilateral upper limb motor imagery tasks for stroke rehabilitation training, *Medicine in Novel Technology and Devices*, vol. 13, no. 100100, 2022, 10.1016/j.medntd.2021.100100.

[17] V. Jeyakumar, P. T. Krishnan, P. Sundaram, and A. N. J. Raj, Chapter 9 – Brain–computer interface in the Internet of Things environment, Editor(s): A. K. Bhoi, V. H. Costa de Albuquerque, S. Nath Sur, and P. Barsocchi (eds.), *Intelligent Data-Centric Systems, 5G IoT and Edge Computing for Smart Healthcare*, Academic Press, pp. 231–255, 2022, ISBN 9780323905480, https://doi.org/10.1016/B978-0-323-90548-0.00012-7.

[18] N. S. Malan and S. Sharma, Motor imagery EEG spectral-spatial feature optimization using dual-tree complex wavelet and neighbourhood component analysis, *IRBM*, vol. 43, no. 3, pp. 198–209, 2022, DOI:10.1016/j.irbm.2021.01.002.

[19] M. A. Cervera, S. R. Soekadar, J. Ushiba, *et al.*, Brain–computer interfaces for post-stroke motor rehabilitation: A meta-analysis, *Annals of Clinical and Translational Neurology*, vol. 5, no. 5, pp. 651–663, 2018, DOI:10.1002/acn3.544.

[20] S. Kulshrestha, M. Agrawal, A. K. Singh, and D. Kulshreshtha, Post-stroke rehabilitation using computer-based cognitive intervention (CBCI): A systematic review, *Current Psychiatry Research, and Reviews*, vol. 16, no. 2, pp. 93–102, 2020, DOI:10.2174/2666082216999200622135105.

[21] J. I. Olszewska, M. Barreto, M. Bermejo-Alonso, *et al.*, Ontology for autonomous robotics, *2017 26th IEEE International Symposium on Robot and Human Interactive Communication (RO-MAN)*, pp. 189–194, 2017, DOI:10.1109/ROMAN.2017.8172300.

[22] F. Parietti, K. C. Chan, B. Hunter, and H. H. Asada, Design and control of supernumerary robotic limbs for balance augmentation, *2015 IEEE International Conference on Robotics and Automation (ICRA)*, pp. 5010–5017, 2015, DOI:10.1109/ICRA.2015.7139896.

[23] Y. Fu, X. Xiong, C. Jiang, B. Xu, Y. Li, and H. Li, Imagined hand clenching force and speed modulate brain activity and are classified by NIRS combined with EEG. *IEEE Transactions on Neural Systems and Rehabilitation Engineering*, vol. 25, no. 9, pp. 1641–1652, 2017, DOI:10.1109/TNSRE.2016.2627809.

[24] C. M. Stinear, C. E. Lang, S. Zeiler, and W. D. Byblow, Advances and challenges in stroke rehabilitation, *Lancet Neurology*, vol. 19, no. 4, pp. 348–360, 2020, DOI: 10.1016/S1474-4422(19)30415-6.

[25] J. Ye, Z. Lu, C. Chen, and M. Wang, Power analysis of a single degree of freedom (DOF) vibration energy harvesting system considering controlled linear electric machines, *2017 IEEE Transportation Electrification Conference and Expo (ITEC)*, pp. 158–163, 2017, DOI:10.1109/ITEC.2017.7993264.

Chapter 5

Comparative analysis of CNN and RNN-LSTM model-based depression detection using modified spectral and acoustic features

Abhishek Ankumnal Matt[1], Yashika Kudesia[1], R. Subashini[1], M. Shaima[1], Abhijit Bhowmick[1], Rajesh Kumar M.[1], Rani C.[2], Aya Hassouneh[3], Lavanya Murugan[4] and M. Murugappan[5,6]

The escalating number of suicides, particularly in India, highlights the importance of addressing depression, anxiety, and stress at the global level. There are several research works reported in the literature that utilize speech signals to detect depression. In this work, a high level of accuracy in depression detection through speech signals is achieved by utilizing two distinct deep neural networks. First, Mel-frequency cepstral coefficients (MFCCs) and fundamental frequencies (F0s) were extracted, where a robust power compression block enhanced the efficiency of MFCC extraction. Furthermore, F0 was calculated based on refined voice samples. In addition, we also extracted other speech-specific features such as spectral flux, spectral centroid, and spectral bandwidth. A feature fusion method was then employed to combine MFCC, spectral flux, spectral centroid, and spectral bandwidth. We used the Distress Analysis Interview Corpus/Wizard-of-Oz (DIAC-WOZ) dataset and used spectral subtraction and the Wiener filter to reduce noise and elevate data quality. In both the recurrent neural network-long short-term memory (RNN-LSTM) and convolutional neural network (CNN) models, a comprehensive evaluation over ten epochs revealed compelling results. The CNN model, however, performed significantly better in the comparative analysis, showing higher F-score (89.0%), precision

[1]School of Electronics Engineering, Vellore Institute of Technology, Vellore, India
[2]School of Electrical Engineering, Vellore Institute of Technology, Vellore, India
[3]Department of Electrical and Computer Engineering, Western Michigan University, USA
[4]School of Engineering and Computing, American International University, Kuwait
[5]Intelligent Signal Processing (ISP) Research Lab, Department of Electronics and Communication Engineering, Kuwait College of Science and Technology, Kuwait
[6]Department of Electronics and Communication Engineering, Faculty of Engineering, Vels Institute of Sciences, Technology, and Advanced Studies, India

(85.0%), recall (76.0%), and overall accuracy (92.31%) compared to the RNN-LSTM model. An acoustic feature-based CNN-based depression detection method could be highly effective for detecting depression at an early stage and could be beneficial for physicians in clinical diagnosis.

Keywords: Depression detection; RNN-LSTM; CNN; Feature fusion; MFCC

5.1   Introduction

It is critical to address the pervasive global problem of depression, which has reached 450 suicides per day in India in 2021. The increase in suicides by 7.2% from the previous year highlights the need for early depression detection. Furthermore, the World Health Organization (WHO) reports that over 700,000 people die by suicide each year. The severity of this public health issue underscores the urgency of addressing it [1]. Affective computing is one of the most powerful methods for detecting depression. The goal of affective computing is to identify, understand, and react to human emotions through technology that is responsive to them [2]. This field combines knowledge from the social sciences, cognitive sciences, and artificial intelligence. Additionally, it plays a significant role in identifying people who are dealing with mental health issues in the context of depression [3]. In addition, affective computing can be utilized to assess depression symptoms through the analysis of data sources such as audio signals (speech), facial expressions, biosignals, and social media text [4]. Consequently, it may be used efficiently for understanding emotional patterns related to depression. These patterns can then be utilized to provide timely support to individuals in need. They may be used to prevent depression's dangerous effects when it is left untreated [5,6].

It has been noted that more research has been conducted on depression detection in recent years, particularly in nations like South Korea where the prevalence of major depression disorders exceeds 40% [7]. In contrast, lower-middle-class families may not be able to pay for the therapy treatments required by traditional depression screening methods. There is a pressing need for a reliable and cost-effective approach to detecting depression. The use of deep learning (DL) techniques to create artificially intelligent clinical diagnosis systems has recently gained increasing interest among researchers. The advantage of this technique is that it can build complex representations from raw data rather than machine learning (ML), which has problems with interpretability and limitations in generalization [8]. In depression detection, recurrent neural networks (RNNs), long short-term memory (LSTM), RNN-LSTM, and convolutional neural networks (CNNs) are examples of the most commonly used DL models in the literature [9]. An RNN-LSTM is a highly efficient way to analyze time-series data. The method is therefore effective at identifying sequential patterns in mood-related data [9]. Furthermore, CNN is capable of identifying spatial patterns in a wide variety of input types, such as text or images. In turn, this improves depression pattern identification sensitivity and accuracy [10].

Natural language processing technologies can analyze textual data and extract emotions, and grammatical patterns from sources such as social media posts to provide valuable insights into users' mental states [11]. Furthermore, graph-based models offer a better understanding of people's mental health in social circumstances by highlighting relationships between data pieces [12]. However, existing depression detection ML and DL methods are subject to biases resulting from imbalanced training data and lack of explainability. Moreover, the collection of large-scale mental health datasets is limited by ethical and privacy concerns [13]. Also, Mel frequency cepstral coefficients (MFCCs), regarded as the top audio characteristics in speech-based applications, have shown significant performance in shallow-based algorithms for depression detection and improved performance when compared to other audio features [14,15].

This study introduces a comprehensive comparison of RNN-LSTM and CNN for depression detection. Our research focused on the key components of fundamental frequency and MFCC using RNN-LSTM. The goal of this novel method is to strengthen MFCCs and improve their effectiveness in detecting depression. Thus, the CNN model was fed with fusion features inputs from MFCC, spectral flux, spectral centroid, and spectral bandwidth to enhance DNN accuracy. As a result of combining these unique advantages, the model is better able to detect subtle depression patterns. Additionally, the current RNN-LSTM and CNN depression detection models were compared and improved for accuracy and performance. Moreover, the robust power compression block for RNN-LSTM and the feature fusion strategy for CNN have been introduced as innovative enhancements.

A critical need for early depression detection is illustrated by this research. Section 5.2 presents the state-of-the-art works related to depression detection using speech signals. Section 5.3 describes our methodology, including feature extraction, fusion, and preprocessing. In Section 5.4, we demonstrate the comparative performance of the RNN-LSTM and CNN models, acknowledge limitations, and suggest future research directions. Finally, Section 5.5 summarizes the research results in the conclusion section.

5.2   State-of-the-art works

The detection of depression is closely associated with stress and emotional stress. Physiological signals have been used to study human emotional stress changes in many studies. In comparison to less accurate methods (such as questionnaires), Bong *et al.* explore methods and approaches for considering human emotional stress changes through multiple physiological signals [16]. They have investigated different types of signals for stress-level classification, including electromyogram (EMG) signals [17] and electrocardiogram (ECG) signals [18]. In addition, researchers have also used ECG signals, EMG signals, and heart rate variability for mental stress assessment [19–21]. Recently, discrete wavelet transform and MFCC features are also used for detecting mental and emotional stress [22,23].

Researchers have investigated a variety of approaches to comprehending auditory symptoms associated with depression through the study of traditional speech characteristics (pitch), intensity, and speech rate [24,25]. There are several features, including MFCCs and fundamental frequencies (F0), that have been proven effective in assessing depression in individuals [14,15]. Moreover, RNNs have a significant ability to learn long-term patterns by capturing sequential dependencies in speech data [26].

In this study, we introduce a novel multimodal RNN approach based on modified MFCCs and F0 fusion to improve depression detection rates. We combine MFCCs, F0, and RNNs to enable more accurate detection of depression using speech signals based on the Distress Analysis Interview Corpus/Wizard-of-Oz (DAIC-WOZ) dataset. About 300 million people are suffering from depression and have limited access to mental health services, according to Xiaolin *et al.* [27]. To overcome obstacles caused by a shortage of psychiatrists and high costs, an accurate depression diagnosis is crucial. They improved recognition accuracy by combining conventional speech features with higher-order spectral features using a fusion feature that uses higher-order spectral analysis.

Our proposed model employs MFCC features and a CNN for emotion classification, demonstrating the superiority of DL for improving speech emotion classification accuracy. The study also introduces DepAudionet, a CNN-based network incorporating fully connected layers, LSTM, and one-dimensional CNN (1D-CNN), serving as a reference for our research. In the research, ensemble methods are explored, which combine multiple models and have shown success in diverse domains, from sentiment analysis to image classification to acoustic environment classification. Researchers in the visual domain focus on modeling facial expressions, head movements, eye gazes, and blinks from facial video data, using representations such as the facial action coding system (FACS) and active appearance model (AAM). A multimodal system combining auditory and visual cues is more effective than one that combines only auditory or visual cues. This research integrates insights from these methodologies to enhance depression detection and understanding through speech.

This study, as described in [28], brings to light a significant gap that exists within the current body of literature pertaining to the management of raw recordings in real-world scenarios. This particular gap focuses on the difficulties related to transmission and storage. Our suggested technique aims to close this gap by addressing the important problem of noise resistance in the features of MFCC. In order to do this, we provide a specially made strong power compression block that is meant to improve the MFCC features' resistance against noise. Our goal in accomplishing this is to guarantee the system's dependability even when there are different noises coming from outside.

Moreover, we have found a broad use of MFCC features in the emotion detection sector by our review of previous studies. However, it is vital to remember that not much research has looked at the critical component of making MFCC features noise-resistant. Furthermore, under the suggested paradigm, only a few research works have carried out a comparative investigation of depression detection utilizing CNN and RNN. With the implementation of a strong power compression block, our work aims to contribute to this context by lowering the noise susceptibility of MFCC features.

Our suggested methodology has shown to be more effective in obtaining spectral characteristics than those that were acquired using earlier techniques. The findings of our investigation demonstrate how well our model can determine whether a person is depressed. It is noteworthy that we want to explore the process of hyperparameter tuning in greater detail in the next versions of this research. By doing this, we want to present a thorough analysis of parameters that will provide insightful information about how they affect the model's overall performance. As said in [29], the goal of this strategy is to improve comprehension of the optimization techniques applied in the proposed model.

5.3 Materials and methods

In this research, we aim to conduct a comparative analysis of two distinct neural networks for depression detection using the DAIC-WOZ dataset. A standard collection comprising 43 samples inclusive of both depressed and non-depressed individuals' audio recordings. To isolate the individual's voice and diminish noise, we perform audio trimming and employ advanced Python libraries, specifically *noisereduce* and *audiofile*, for preprocessing. The enhanced audio samples undergo feature extraction in the two approaches.

In our initial method, we incorporated MFCC as a representation of the short-term power spectrum of a sound combined with fundamental frequency as a crucial feature. The calculation of MFCC involves using (5.1) [30]:

$$\text{MFCC}_i = \sum_{k=1}^{N} X_k \cos \cos \left[i \left(k - \frac{1}{2} \right) \frac{\pi}{2} \right], i = 1, 2, 3, \ldots, M \tag{5.1}$$

where N is the number of MFCC coefficients to be computed, i is the index of the resulting MFCC, M is the number of cepstrum coefficients, and X_k, $K = 1, 2, \ldots, N$, represents the log-energy output of the kth filter.

We modify it by adding a robust compression block to the fundamental framework to address MFCC's sensitivity to noise, ensuring its suitability for real-time applications. These features, after undergoing this modification, are then processed through an RNN-LSTM model, and we meticulously document the corresponding performance parameters.

In our second approach, we expanded our feature set to encompass a wider array, incorporating MFCC alongside spectral flux (5.4) to measure the change in the power spectrum of a signal between consecutive frames [31], spectral centroid (5.2) to identify the center of mass of the spectrum and provide information about the "brightness" of the sound [31], and spectral bandwidth (5.3) to measure the width of the frequency band occupied by a signal and indicate the spread of frequencies [31], as shown in Figure 5.1. An enhancement of accuracy was achieved by combining the conventional MFCC extraction method with a feature fusion

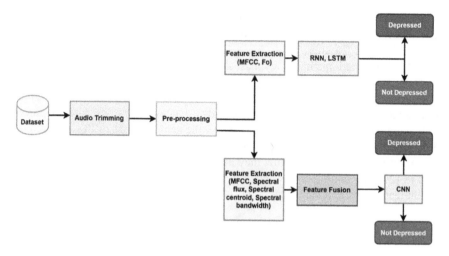

Figure 5.1   Proposed framework

technique. As a result, the fused features were fed into the CNN model, and the performance parameters were carefully recorded. Lastly, the study compared the performance parameters of both approaches to determine which model was more accurate in detecting depression.

5.3.1   Dataset

This study used the DIAC-WOZ dataset [32]. This dataset consists of a comprehensive collection of clinical interviews (audio and video recordings, along with extensive questionnaire responses) designed to diagnose psychological distress conditions such as anxiety, depression, and post-traumatic stress disorder. We developed a computer agent based on these interviews that could conduct interviews and recognize verbal and non-verbal indicators of mental illness. The audio recordings were made using Sennheiser HSP 4-EW-3 head-mounted microphones at a 16 kHz sampling rate. Although there may be minimal bleed-over of the virtual interviewer in the audio files, we recommend relying on the transcript files to address this concern. A portion of the audio recordings has been scrubbed to ensure privacy when utterances could identify individuals. In these cases, the waveform is zeroed out, and the transcript files have the keyword "scrubbed_entry" to facilitate identification. To ensure consistency and privacy, entries that are flagged as scrubbed are also zeroed out in their corresponding feature files. Table 5.1 provides a detailed description of the dataset.

5.3.2   Preprocessing

Preprocessing of audio samples is an essential part of the modeling process because it has a big impact on the model's accuracy and the settings that are needed for robust and reliable results over time. A reduction in the effects of noise and an

Table 5.1   DIAC-WOZ dataset characteristics

Characteristic	Details
No. of subjects	189 clinical interviews (30/107 within the training set and 12/35 within the development set are classified as depressed)
No. of video recordings	DAIC-WOZ comprises over 50 h of speech recordings, aligning with just 189 individual samples
No. of trials	Training, validation, and test sets are 60%, 20%, and 20%, respectively
No. of classes	The experiments involve two classes: "has depression/no depression" and a separate two-class classification for depression severity
Sampling frequency	16 kHz
Operating environment	A patient interacts with a digital avatar controlled remotely. Using this avatar, a clinician asks targeted questions to pinpoint depressive symptoms. The audio was captured at 16 kHz via a head-mounted microphone, while video recorded at 30 frames per second with a Microsoft Kinect captured a total of 68 facial key points in three dimensions using OpenFace
Gender ratio	Not specified
Age	Not provided
Other factors	• For each participant, DAIC-WOZ provides a Patient Health Questionnaire (PHQ-8) score, which indicates the depression severity • A binary label of the PHQ-8 score is provided to represent the presence of depression. A score ≥ 10 indicates that the participant is suffering from depression

improvement in the quality of speech signals is necessary to enhance speech quality. We used advanced noise reduction techniques (such as the *"noise reduce"* Python library) to refine our audio samples and enhance the overall quality of the audio to achieve this goal.

5.3.3   Depression detection using RNN-LSTM

5.3.3.1   Feature extraction

Modified spectral feature

As shown in Figure 5.2, a series of sequential steps is employed to refine audio signals through MFCC extraction. First, high-frequency components are enhanced, and then the audio stream is divided. As a next step, the signal is converted into the frequency domain using the fast Fourier transform after minimizing spectral leakage using a Hamming window. After that, the Mel filter bank is used to record the distribution of energy across frequency bands. By utilizing power compression and a logarithm operation, characteristics are enhanced against loudness changes. Finally, a discrete cosine transform is used to extract MFCCs and reduce dimensionality.

Fundamental frequency

This process estimates the fundamental frequency and gathers relevant voicing information using the Python function *"librosa.piptrack."* The F0 (the frequency at which the vocal cords vibrate during speech) is extracted from the pitch tracker,

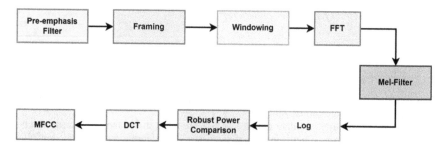

Figure 5.2   Proposed MFCC feature extraction framework

especially in frames with signs of voicing. This yields a quantitative measure of the fundamental frequency for each audio file by calculating the mean F0. It is important to identify changes in speech patterns, prosody, and intonation associated with emotional health using F0. A computational model designed to detect depression must therefore include F0 extraction as an essential feature. Table 5.2 illustrates three features for 43 audio samples: MFCC mean, MFCC variance, and fundamental frequency. Additionally, it indicates the depression status of each sample (Labels), with "1" indicating depression and "0" indicating the absence of depression.

5.3.3.2   Framework of the depression detection

The presented methodology successfully distinguishes individuals in a healthy state from those with depression using features extracted from their speech signals, typically derived from narrative speech segments that describe the subjects' daily activities. Figure 5.3 illustrates the block diagram of our approach used to detect the depression using RNN-LSTM.

5.3.3.3   RNN-LSTM model

It is important to note that the suitability of a model depends on many factors, including the characteristics of the dataset, task demands, and computational capacity; however, the preference for RNN-LSTM architectures is largely determined by their ability to manage sequential data. Since they are effective at learning long-range sequences, depression-related features in speech may extend over multiple time steps, so they were selected for depression detection. Furthermore, they are capable of handling variable input lengths, learning relevant features automatically, and fitting time series models. Figure 5.4 illustrates the standard scalar procedure for training a model on a labeled dataset with a standard deviation of 1 and zero mean.

Following this, essential sequences are created for the training, validation, and test sets, a crucial preparatory step for training an RNN. The RNN model, featuring LSTM layers and fully connected layers, is compiled using the Adam optimizer and binary cross-entropy loss. The training process incorporates early stopping and model checkpointing mechanisms to mitigate overfitting. Subsequently, the model

Table 5.2   Extracted features from the enhanced samples

Subject no.	MFCC mean	MFCC VAR	F0 mean	Labels
303	69.829994	935307.5643	834.19	0
304	66.606969	799844.4614	1080.13	1
305	66.817799	933800.8484	872.27	1
310	411.179907	10611191.12	920.23	1
312	446.512569	11623298.16	759.22	0
313	719.81283	18760336.2	686.45	1
315	231.753839	5617856.911	1071.38	0
316	587.198991	14927250.99	945.26	1
317	655.647803	17216921.15	942.91	1
318	676.910345	17736747.63	776.44	0
319	340.701918	8447423.185	1205.62	1
320	735.236077	18876889.79	981.33	1
321	310.154941	7348303.153	1017.02	0
322	476.741931	12306645.9	769.36	1
324	238.766285	5859283.411	1179.3	1
325	260.135364	6149701.641	1034.6	1
326	403.126503	10435789.45	831.41	0
327	81.12397	1044306.601	1091.94	0
328	270.355347	6710463.176	1197.63	0
330	894.286553	22920695.43	948.43	1
336	674.303549	17326407.23	1051.33	1
300	56.125636	39523.94085	1228.4	0
301	325.582372	8291029.058	1002.52	1
306	57.626525	892954.4063	918.5	0
311	683.730024	17461692.45	1140.59	1
323	454.475376	11235670.18	907.77	0
329	259.603411	6283509.679	812.53	0
331	243.232395	6020734.07	1120.06	1
332	960.564356	25138387.2	1283.74	1
334	162.630235	3825092.827	998.9	1
335	304.385835	7384278.8	948.65	1
337	418.286658	10483967.72	1300.63	1
346	789.361173	20361999.35	1299.14	1
354	409.08364	10124438.82	1181.81	1
359	219.223001	5388705.553	1146.47	1
361	134.289801	2806708.387	951.39	0
365	392.469038	9862081.598	1162.48	1
378	373.583233	9164091.601	1104.15	1
384	998.508897	25873523.25	899.71	1
387	1056.502862	27386989.05	1210.88	0
399	256.556048	5641842.719	1112.95	1
407	333.448982	8497950.112	650.42	0

is evaluated on the test set, presenting metrics such as loss Figure 5.5(a) and accuracy Figure 5.5(b).

Another aspect introduces hyperparameter tuning through GridSearchCV, optimizing critical parameters like the activation function, dropout rate, and

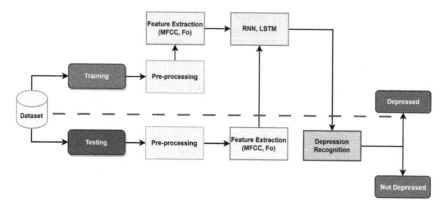

Figure 5.3   Overall workflow procedure of speech depression classification

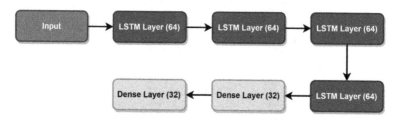

Figure 5.4   Block diagram illustrates the flow of data through the RNN model

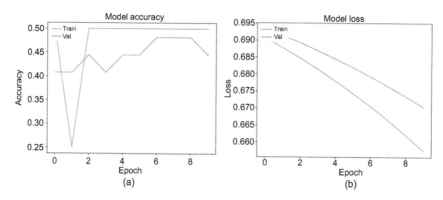

Figure 5.5   The plot illustrates the training and validation accuracy (a) and loss (b) over epochs

optimizer. The best model derived from this process undergoes evaluation on the test set. Finally, an illustrative example (shown in Figure 5.6) involves generating synthetic data for classification, subjecting it to hyperparameter tuning, and evaluating its performance, thereby highlighting the adaptability of the approach.

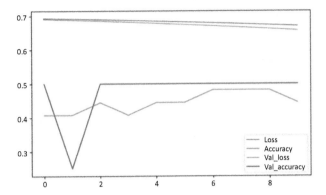

Figure 5.6   The DataFrame plot showcases the training and validation metrics (accuracy and loss)

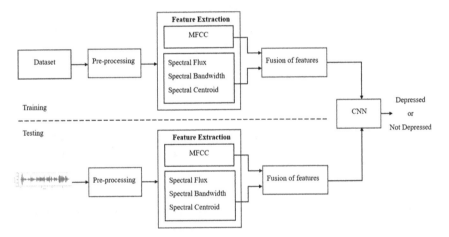

Figure 5.7   Overall workflow procedure of speech depression classification

5.3.4   Depression detection using CNN

5.3.4.1   Proposed framework

Figure 5.7 presents the comprehensive workflow for speech-based depression classification. Beginning with dataset preprocessing, it culminates in determining whether an individual exhibits signs of depression or not.

5.3.4.2   Feature extraction

Mel frequency cepstral coefficients

The Python library librosa facilitates the extracting of MFCC values from audio samples [33]. Figure 5.8 illustrates MFCC plots for depressed and non-depressed audio samples. These plots reveal that, for non-depressed audio samples, speech intensity is lower at higher frequencies and more concentrated at lower frequencies.

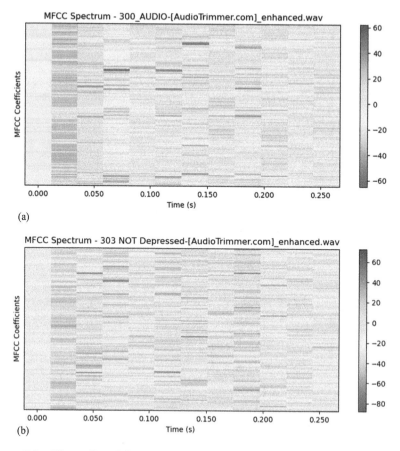

Figure 5.8   Plots of MFCC: (a) depressed audio sample and (b) non-depressed audio sample

In contrast, depressed audio samples exhibit high-frequency components with greater intensities, especially during brief time durations.

The second feature set comprises acoustic features such as spectral centroid, spectral bandwidth, and spectral flux, each playing a unique role in capturing essential elements of the audio stream. The spectral centroid reflects the brightness or pitch of the sound—higher values denote brighter or higher-pitched sounds, while lower values indicate duller or lower-pitched ones. Spectral flux assesses the change in spectral content over time within the audio signal; higher values indicate a more dynamic spectrum, while lower values indicate greater stability.

Spectral centroid

The spectral centroid represents the center of the spectrum or the "gravity center." To compute the spectral centroid, each frame in the magnitude spectrogram is normalized to create a distribution over frequency bins. The centroid of each frame at time "t" is found by summing up all frequency values in the distribution and weighting them according to magnitude values. This measure highlights regions of a frame containing the bulk of the spectral energy. The spectral centroid is obtained using the Python function librosa.feature.spectral_centroid( ) (Table 5.3) [33].

Table 5.3 Features extracted from the enhanced samples

Subject no.	MFCC mean	Spectral flux mean	Spectral centroid mean	Spectral bandwidth mean
1	−24.3747	0.96889	1740.146	1962.812906
2	−19.2786	1.342574	1356.598	1518.601951
3	−23.2689	1.556813	1331.263	1528.954441
4	−23.5972	1.477168	1232.356	1376.416004
5	−7.85222	1.038695	726.4961	942.3269193
6	−11.0767	1.214975	905.0743	1159.953799
7	−7.91541	1.290629	964.9831	1098.045688
8	−19.501	1.677847	1179.573	1440.577014
9	−33.6733	1.559532	1721.349	1792.334525
10	−16.8277	1.618247	1027.935	1303.351371
11	−15.6604	1.58381	1086.708	1339.281951
12	−30.8401	1.216235	1122.246	1595.756488
13	−24.4895	1.290861	1130.175	1384.978581
14	−30.1621	1.509766	1377.344	1530.984513
15	−31.7414	1.275353	1295.433	1684.778714
16	−34.1597	1.763976	1391.474	1482.327485
17	−28.938	1.725456	1131.192	1404.204933
18	−34.4771	1.729034	1534.197	1759.054991
19	−30.1039	2.041832	1385.081	1446.303423
20	−24.3584	1.736742	1141.138	1465.708695
21	−32.3415	1.877218	1470.343	1410.539504
22	−26.0832	1.8032	1401.158	1404.989922
23	−26.2748	1.750097	1303.763	1422.120953
24	−34.5478	1.226871	1436.666	1728.679216
25	−24.7035	1.928086	1541.433	1490.267192
26	−25.4033	1.708282	1422.081	1405.130841
27	−30.3461	1.15814	1387.651	1753.411678
28	−22.86	1.966397	1273.733	1461.085464
29	−30.9435	1.979677	1532.451	1500.756232
30	−32.0184	1.7776	1565.99	1618.411154
31	−26.4063	2.156775	1344.262	1355.487143
32	−28.4084	1.472469	1298.299	1558.145132
33	−34.5011	1.411404	1364.763	1626.198743
34	−34.5011	1.411404	1364.763	1626.198743
35	−32.6785	1.326198	1257.367	1501.705207
36	−23.8873	1.931277	1258.596	1381.641122
37	−29.2835	1.779139	1462.642	1425.184282
38	−27.8328	1.574865	1210.623	1543.372892
39	−31.2511	1.85284	1438.291	1478.341741
40	−34.1121	1.218937	1350.514	1636.508669
41	−33.4462	0.761884	1363.642	1780.006594
42	−35.5014	1.379547	1238.003	1606.578318
43	−25.5164	1.793816	1211.562	1381.486709

The computation of spectral centroid is calculated using (5.2) [31]:

$$\text{Spectral Centroid} = \frac{\sum_{k=b1}^{b2} f_k S_k}{\sum_{k=b1}^{b2} S_k} \tag{5.2}$$

Here, f_k is the frequency bin, and S_k is the spectral magnitude at frequency f_k in frame n.

Spectral bandwidth
Spectral bandwidth, the difference between the higher and lower frequencies in a continuous spectrum, is computed as the sum of the largest deviations of the signal on either side of the centroid. Librosa calculates spectral bandwidth using (5.3):

$$\sum_k^t S[k,t] \times (\textit{freq}[k,t] - \text{centroid}[t]) \times p \times \frac{1}{p} \tag{5.3}$$

where "p" represents the order, and "t" represents time. The spectral bandwidth is obtained using the Python function "*librosa.feature.spectral_bandwidth( )*" [33].

Spectral flux
Spectral flux assesses how quickly the spectral content of an audio signal changes between successive frames. It calculates the rate of change of the frequency domain energy distribution over time. Spectral flux is computed as the squared difference between the spectral magnitudes of the previous frame and the current frame, averaged over all frequency bins. This metric reflects the dynamic nature of an audio stream. Librosa calculates spectral flux using the "librosa.onset.onset_strength()" function. The envelope of spectral flux onset strength is determined by (5.4):

$$\text{mean}_f (0, \text{ref}[f, t - \text{lag}] - S[f, t]) \tag{5.4}$$

where following local maximum filtering along frequency axis 1, "ref" is "S." If time series "y" is provided, "S" will be the default. A combination of MFCC analysis and acoustic features is used to detect mental health issues, especially depression [33].

5.3.4.3   Feature fusion
The fusion of features is an important stage for understanding the basic characteristics that affect depression detection later in the ML process. Several functions and procedures are involved in the fusion process in Python. As a result of the feature extraction functions, such as *extract_mfcc*, *calculate_spectral_flux*, *calculate_spectral_centroid*, and *calculate_spectral_bandwidth*, distinct audio features are computed, including mean MFCC, spectral flux, spectral centroid, and spectral bandwidth. Feature fusion lies at the core of the *process_and_plot function*, which not only extracts features but represents them graphically as well. By combining the mean values of each feature into one array named "*fused_features*," the main loop invokes the "*process_and_plot*" function. Afterward, this array of fused features is appended to a list, while the corresponding labels ("depressed" or "not_depressed") are appended to another list. The cumulative result is then

encapsulated in a DataFrame, df_fused, where each row represents an audio sample, characterized by its fused features and labels. Lastly, this DataFrame serves as the basis for detecting depression patterns within the aggregated audio data.

5.3.4.4 CNN model

A CNN model is used to classify audio samples into two categories: "Class 0" for non-depressed individuals and "Class 1" for depressed individuals. A preprocessed and encoded labeled audio feature is first loaded into the system. Following that, the dataset is divided into training sets and testing sets. Then, the features are standardized, so the CNN can scale them consistently, and then the features are reshaped so the CNN can process them. Labels undergo one-hot encoding, which is a standard method for classifying data. CNN models are built using Keras using key layers such as 1D convolutional layers, max-pooling layers, flattening layers, and dense hidden layers activated by rectified linear units.

The output layer uses softmax activation for multiclass classification. The model is compiled using categorical cross-entropy as the loss function, the Adam optimizer, and accuracy as the evaluation metric. Subsequently, the model undergoes training for 20 epochs with a batch size of 4, and the process is monitored using validation data. After training completion, the model is evaluated on the test set, and the achieved accuracy is reported. Predictions are then generated for the test set, and key classification metrics, including precision, recall, and F1-score, are computed and presented in a classification report. The overarching goal is to leverage the CNN's capability to capture intricate patterns within the fused audio features, facilitating the identification of depression based on learned patterns.

5.4 Numerical results and discussion

5.4.1 Results of RNN-LSTM model

Confusion matrix
The confusion matrix (Table 5.4) relies on these Equations (5.5)–(5.8):

$$\text{Accuracy} = \frac{\text{TP} + \text{TN}}{\text{TP} + \text{TN} + \text{FP} + \text{FN}} \tag{5.5}$$

$$\text{Precision} = \frac{\text{TP}}{\text{TP} + \text{FP}} \tag{5.6}$$

$$\text{Recall} = \frac{\text{TP}}{\text{TP} + \text{FN}} \tag{5.7}$$

$$F1 - \text{Score} = 2 * \frac{\text{Precision} * \text{Recall}}{\text{Precision} + \text{Recall}} \tag{5.8}$$

where

- TP (True Positive) represents the count of accurately predicted positive instances.

Table 5.4   Confusion matrix for the RNN-LSTM

Actual values	Predicted values	
	84	9
	25	82
Test accuracy		0.83
Test precision		0.9
Test recall		0.76
Test F1-score		0.82

Notes:
- The top left bold value of 84 represents the number of instances that were actually negative (not depressed) and were correctly predicted as negative by the model (true negatives).
- The top right bold value of 9 represents the number of instances that were actually negative but were incorrectly predicted as positive (depressed) by the model (false positives).
- The bottom left bold value of 25 represents the number of instances that were actually positive (depressed) but were incorrectly predicted as negative by the model (false negatives).
- The bottom right bold value of 82 represents the number of instances that were actually positive and were correctly predicted as positive by the model (true positives).
- The bold values in the confusion matrix provide insights into the performance of the RNN-LSTM model by showing how well it classified instances into the correct categories (depressed or not depressed). Higher values in the top-left and bottom-right cells indicate better model performance, while higher values in the top-right and bottom-left cells indicate more misclassifications by the model.

- TN (True Negatives) denotes the number of accurately predicted negative instances.
- FP (False Positives) signifies the count of incorrectly predicted positive instances.
- FN (False Negatives) stands for the number of incorrectly predicted negative instances.

5.4.2   Results of the CNN model

The effectiveness of a binary classification model is assessed in the classification report. Metrics like accuracy, recall, and F1-score for non-depressed sample are all 0.00, reflecting that there is just one case in the dataset and no accurate positive predictions. The depressed sample, on the other hand, performs admirably, with an F1-score of 0.96, recall of 1.00, and accuracy of 0.92. For accuracy, recall, and F1-score, the weighted average (taking imbalances into account) is 0.85, 0.92, and 0.89, whereas the macro average is 0.46, 0.50, and 0.48 across classes. Although the model's accuracy of 92.31% is noteworthy, there are noticeable difficulties in accurately predicting non-depressed sample cases. This thorough assessment identifies the model's advantages as well as limitations, providing valuable insights.

In comparing the performance of RNN-LSTM and CNN classifiers, notable differences emerge. The F1-score, a metric that balances precision and recall, indicates that the CNN classifier achieves a higher score of 0.89 compared to the RNN-LSTM model's score of 0.82. This suggests that the CNN algorithm excels in capturing a balance between true positives, false positives, and false negatives. Looking at precision, the RNN-LSTM model demonstrates a higher precision of 0.9 compared to the CNN model's precision of 0.85. Conversely, the CNN model outperforms the RNN-LSTM model in the recall, securing a higher score of 0.92 compared to the RNN-LSTM model's achievement of 0.76. The CNN model outperforms the RNN-LSTM model by achieving an accuracy of 92.31% compared to 83.55% for the RNN-LSTM model. It can be concluded from these results that the RNN-LSTM algorithm is better in terms of precision, whereas the CNN algorithm is better in terms of F1-score, recall, and overall accuracy than the RNN-LSTM algorithm (Table 5.5).

Our test results indicate that both the RNN-LSTM and CNN models are viable for depression detection (Table 5.6). However, the outcomes unmistakably favor the CNN model, signifying its superiority in performance. This distinction is a compelling rationale for our exploration of a second method involving feature fusion. As established, model performance tends to improve when various features are fused. This underscores the critical role of feature fusion in enhancing results and ensuring an increase in model accuracy. Consequently, in our subsequent

Table 5.5   Model's output parameters

Actual values	Predicted values	
	0.6	0.2
	1.1	20.1
Test accuracy		0.94
Test precision		0.85
Test recall		0.92
Test F1-score		0.89

Table 5.6   Comparative analysis of RNN-LSTM and CNN

Performance measures	RNN-LSTM	CNN
F1-score	0.82	0.89
Precision	0.90	0.85
Recall	0.76	0.92
Accuracy	0.83	0.94

approach, the CNN model maintains a distinct advantage over the RNN-LSTM model. The inclination toward CNN is reinforced by the belief that feature fusion acts as a key determinant in achieving superior results. Still, the outcomes clearly show how much better the CNN model performs.

5.4.3   Major limitations

- **Imbalanced dataset:** The reported difficulties in accurately predicting non-depressed sample cases may be attributed to the imbalanced nature of the dataset. With just one case in the depressed sample, the model's ability to generalize to instances without the specific characteristic is limited.
- **Limited generalizability:** The results of this study are based on a specific dataset (DIAC-WOZ). Consequently, the model may not be generalizable.
- **Dependence on features:** The comparison between RNN-LSTM and CNN models depends on the features used (MFCC, spectral flux, spectral centroid, and spectral bandwidth). In this sense, the effectiveness of the models may vary with different feature sets, and additional exploration of other features is needed for a more comprehensive understanding.
- **Single metric evaluation:** Multimetric evaluation is preferable, since F1-score, precision, recall, and accuracy are only one metric evaluation, and they may not be sufficient to explain model performance in its entirety.

5.4.4   Future work

- **Dataset expansion:** To improve model generalization and enhance depression detection system robustness, future work must include a larger and more varied dataset.
- **Fine-tuning model parameters:** In this study, 10 epochs were used for training the model, so tuning hyperparameters (number of epochs, learning rates, and network architectures) helps address overfitting and underfitting issues.
- **Ensemble approaches:** The use of ensemble models that combine RNN-LSTM and CNN strengths could often minimize model weaknesses and improve overall prediction.
- **Real-world deployment considerations:** It is imperative to consider deployment challenges, ethical implications, as well as user acceptance in the real world. Therefore, future research should ensure that depression detection models are user-friendly and aligned with ethical standards.
- **Longitudinal studies:** It may be useful to conduct longitudinal studies for individuals with depression to track changes in speech patterns over time to gain a better understanding of the relationship between speech and mental health.

5.5   Conclusion

In our research, we used the DIAC-WOZ dataset and implemented preprocessing techniques, specifically spectral subtraction, and the Wiener filter, to effectively reduce noise and enhance the overall data quality. To explore the most effective techniques and models for achieving high accuracy, we followed two distinct approaches. In the first approach, key features extracted encompassed MFCC and fundamental frequency (F0). Notably, we introduced a robust power compression block to enhance the efficiency of the MFCC extraction process. Additionally, F0 was derived from the enhanced voice samples. In the second approach, we incorporated feature fusion to further improve accuracy. The extracted features in this approach included MFCC, spectral flux, spectral centroid, and spectral bandwidth. During the training phase, we conducted a comprehensive evaluation with 10 epochs to closely monitor the model's progression. The results from both approaches are presented as follows for the RNN-LSTM and CNN models in tabular column 5. These findings clearly show that the CNN model is superior to the RNN-LSTM model, as evidenced by the CNN model's higher F1-score, precision, recall, and total accuracy. The predictive capacity of the model has been shown to be enhanced by the second approach's use of feature fusion together with robust preprocessing methods. The increased accuracy of CNN indicates that it is an effective approach for the detection of depression.

References

[1] Z. Jalilian, F. Mohamadian, Y. Veisani, and S. Ahmadi, "The trend of death and YLL (Years of Life Lost) of disease due to social harms (Suicide, Homicide and Addiction), Ilam Province, 2009–2019," *Bull Emerg Trauma*, 2023.

[2] O. P. Singh, "Startling suicide statistics in India: Time for urgent action," *Indian Journal of Psychiatry*, vol. 64, no. 5. Medknow, 2022, pp. 431–432.

[3] C. Zucco, B. Calabrese, and M. Cannataro, "Sentiment analysis and affective computing for depression monitoring," in *2017 IEEE International Conference on Bioinformatics and Biomedicine (BIBM)*, IEEE, 2017, pp. 1988–1995.

[4] J. F. Cohn and F. De la Torre, "Automated face analysis for affective computing," in R. Calvo, S. D'Mello, J. Gratch, and A. Kappas (eds), *The Oxford Handbook of Affective Computing*, New York, NY: Oxford University Press, 2015, p. 131.

[5] S. Poria, E. Cambria, R. Bajpai, and A. Hussain, "A review of affective computing: From unimodal analysis to multimodal fusion," *Information Fusion*, vol. 37, 2017, pp. 98–125.

[6] S. Jayawardena, J. Epps, and E. Ambikairajah, "Evaluation measures for depression prediction and affective computing," in *ICASSP 2019-2019 IEEE International Conference on Acoustics, Speech and Signal Processing (ICASSP)*, IEEE, 2019, pp. 6610–6614.

[7] M. M. Ohayon and S.-C. Hong, "Prevalence of major depressive disorder in the general population of South Korea," *Journal of Psychiatry Research*, vol. 40, no. 1, 2006, pp. 30–36.

[8] A. Shrestha and A. Mahmood, "Review of deep learning algorithms and architectures," *IEEE Access*, vol. 7, 2019, pp. 53040–53065.

[9]   A. Amanat, M. Rizwan, A. R. Javed, *et al.*, "Deep learning for depression detection from textual data," *Electronics (Basel)*, vol. 11, no. 5, 2022, p. 676.

[10] Z. Zhao, Z. Yang, L. Luo, *et al.*, "ML-CNN: A novel deep learning based disease named entity recognition architecture," in *2016 IEEE International Conference on Bioinformatics and Biomedicine (BIBM)*, IEEE, 2016, p. 794.

[11] W. Zheng, L. Yan, C. Gou, and F.-Y. Wang, "Graph attention model embedded with multi-modal knowledge for depression detection," in *2020 IEEE International Conference on Multimedia and Expo (ICME)*, IEEE, 2020, pp. 1–6.

[12] S. Alghowinem, "From joyous to clinically depressed: Mood detection using multimodal analysis of a person's appearance and speech," in *2013 Humaine Association Conference on Affective Computing and Intelligent Interaction*, IEEE, 2013, pp. 648–654.

[13] T. Richter, B. Fishbain, G. Richter-Levin, and H. Okon-Singer, "Machine learning-based behavioral diagnostic tools for depression: advances, challenges, and future directions," *Journal of Personalized Medicine*, vol. 11, no. 10, 2021, p. 957.

[14] Z. Zhao, Z. Bao, Z. Zhang, *et al.*, "Automatic assessment of depression from speech via a hierarchical attention transfer network and attention auto-encoders," *IEEE Journal of Selected Topics in Signal Processing*, vol. 14, no. 2, 2019, pp. 423–434.

[15] Y. Wang, L. Liang, Z. Zhang, *et al.*, "Fast and accurate assessment of depression based on voice acoustic features: a cross-sectional and longitudinal study," *Front Psychiatry*, vol. 14, 2023, p. 1195276.

[16] S. Z. Bong, M. Murugappan, and S. Yaacob, "Methods and approaches on inferring human emotional stress changes through physiological signals: A review," *International Journal of Medical Engineering and Informatics*, vol. 5, no. 2, 2013, pp. 152–162.

[17] P. Karthikeyan, M. Murugappan, and S. Yaacob, "EMG signal based human stress level classification using wavelet packet transform," in *Trends in Intelligent Robotics, Automation, and Manufacturing: First International Conference, IRAM 2012, Kuala Lumpur, Malaysia, November 28–30, 2012. Proceedings*, Springer, 2012, pp. 236–243.

[18] S. Z. Bong, M. Murugappan, and S. Yaacob, "Analysis of electrocardiogram (ECG) signals for human emotional stress classification," in *Trends in Intelligent Robotics, Automation, and Manufacturing: First International Conference, IRAM 2012, Kuala Lumpur, Malaysia, November 28–30, 2012. Proceedings*, Springer, 2012, pp. 198–205.

[19] P. Karthikeyan, M. Murugappan, and S. Yaacob, "ECG signals based mental stress assessment using wavelet transform," in *2011 IEEE International Conference on Control System, Computing and Engineering*, IEEE, 2011, pp. 258–262.

[20] B. S. Zheng, M. Murugappan, and S. Yaacob, "Human emotional stress assessment through heart rate detection in a customized protocol experiment," in *2012 IEEE Symposium on Industrial Electronics and Applications*, IEEE, 2012, pp. 293–298.

[21] B. S. Zheng, M. Murugappan, S. Yaacob, and S. Murugappan, "Human emotional stress analysis through time domain electromyogram features," in *2013 IEEE Symposium on Industrial Electronics & Applications*, IEEE, 2013, pp. 172–177.

[22] M. Murugappan, N. Q. I. Baharuddin, and S. Jerritta, "DWT and MFCC based human emotional speech classification using LDA," in *2012 International Conference on Biomedical Engineering (ICoBE)*, IEEE, 2012, pp. 203–206.

[23] P. Karthikeyan, M. Murugappan, and S. Yaacob, "A study on mental arithmetic task based human stress level classification using discrete wavelet transform," in *2012 IEEE Conference on Sustainable Utilization and Development in Engineering and Technology (STUDENT)*, IEEE, 2012, pp. 77–81.

[24] M. Cannizzaro, B. Harel, N. Reilly, P. Chappell, and P. J. Snyder, "Voice acoustical measurement of the severity of major depression," *Brain Cognition*, vol. 56, no. 1, 2004, pp. 30–35.

[25] L. Albuquerque, A. R. S. Valente, A. Teixeira, *et al.*, "Association between acoustic speech features and non-severe levels of anxiety and depression symptoms across lifespan," *PLoS One*, vol. 16, no. 4, 2021, p. e0248842.

[26] B. Yue, J. Fu, and J. Liang, "Residual recurrent neural networks for learning sequential representations," *Information*, vol. 9, no. 3, 2018, p. 56.

[27] X. Miao, Y. Li, M. Wen, Y. Liu, I. N. Julian, and H. Guo, "Fusing features of speech for depression classification based on higher-order spectral analysis," *Speech Communication*, vol. 143, 2022, pp. 46–56.

[28] M. A.-E. A. Ezzi, N. N. W. N. Hashim, N. A. Basri, and S. F. Toha, "Speech-based depression detection for Bahasa Malaysia female speakers using deep learning," *ELEKTRIKA-Journal of Electrical Engineering*, vol. 20, no. 2–3, 2021, pp. 1–6.

[29] A. Vázquez-Romero and A. Gallardo-Antolín, "Automatic detection of depression in speech using ensemble convolutional neural networks," *Entropy*, vol. 22, no. 6, 2020, p. 688.

[30] S. Davis and P. Mermelstein, "Comparison of parametric representations for monosyllabic word recognition in continuously spoken sentences," *IEEE Transactions on Acoustics*, vol. 28, no. 4, 1980, pp. 357–366.

[31] S. A. A. Aleem, M. F. M. Yusof, M. Quazi, M. A. Halil, and M. Ishak, "Mel-frequency cepstral and spectral flux analysis of the acoustic signal for real-time status monitoring of laser cleaning," *Materials Research Express*, vol. 10, no. 10, 2023, p. 106506.

[32] J. Gratch, R. Artstein, G. Lucas, *et al.*, "The distress analysis interview corpus of human and computer interviews," in *LREC*, Reykjavik, 2014, pp. 3123–3128.

[33] B. McFee, C. Raffel, D. Liang, *et al.*, "Librosa: Audio and music signal analysis in python," in *Proceedings of the 14th Python in Science Conference*, 2015, pp. 18–25

Chapter 6

Unraveling emotions: harnessing pre-trained convolutional neural networks for electroencephalogram signal analysis

Md. S. Bin Islam[1,2], Md. S.I. Sumon[1,2], M. Murugappan[3,4,5] and M.E.H. Chowdhury[2]

Convolutional neural networks, also known as CNNs, have shown notable progress in the domain of biological data processing. These strategies have the potential to facilitate the identification of emotions within the context of an emotional brain–computer system. This chapter introduces an innovative methodology for the identification and analysis of emotions. The system utilizes efficient connectivity and finely adjusted CNNs based on a multichannel electroencephalogram (EEG) input. After preprocessing the EEG data, the study investigates the direct directed transfer function (dDTF) technique used to figure out the relationships between the 32 channels of EEG. Following this, the EEG data was used to produce visual representations for every individual, which were then fed into three CNN models that had been pre-trained: Inception-v3, ResNet-50, and VGG-19. The efficacy of the suggested technique is assessed by conducting evaluations on the MAHNOB-HCI and Dataset for Emotion Analysis using Physiological Signals (DEAP) databases. The experimental study included the examination of five different emotional states (neutral and four-quarters of the valence–arousal model). The study found that using ResNet-50 on dDTF pictures in the alpha band frequency spectrum (8–13 Hz) produced the most favorable results. The efficiency with which the brain's connection is captured may be attributed to the specific design of the subject in question. The MAHNOB-HCI system obtained high levels of accuracy and F1-score metrics, with values of 99.41% and 99.42%, respectively. The DEAP databases had

[1]Department of Biomedical Engineering, Military Institute of Science and Technology, Bangladesh
[2]Department of Electrical Engineering, Qatar University, Qatar
[3]Intelligent Signal Processing (ISP) Research Lab, Department of Electronics and Communication Engineering, Kuwait College of Science and Technology, Kuwait
[4]Department of Electronics and Communication Engineering, Faculty of Engineering, Vels Institute of Sciences, Technology, and Advanced Studies, India
[5]Centre for Excellence in Unmanned Aerial Systems (CoEUAS), Universiti Malaysia Perlis, Malaysia

accuracy rates of 98.17% and 98.23%. A novel approach has been proposed to effectively investigate the functioning of the brain.

Keywords: Convolutional neural networks (CNN); Electroencephalogram (EEG); Dynamic directed transfer function (dDTF) method; Physiological signals

6.1 Introduction

The acknowledgment of human affect has great importance in the field of mental health treatment and overall well-being, as it plays a substantial part in resilience, motivation, social impact, and empathy. Affective computing is the field dedicated to developing systems capable of seeing, comprehending, analyzing, and imitating human emotions. Its popularity has increased in recent years. Electroencephalogram (EEG), electromyogram (EMG), electrocardiogram (ECG), and electrodermal activity (EDA) are physiological signals that can provide insights into individuals' emotional states [1]. A major challenge in emotional computing lies in the precise and real-time detection of these signals [2]. For affective state mining to be effective, these indicators are essential as they provide valuable insights into individuals' emotions and thoughts. Traditional methods for recognizing affective states often rely on the manual extraction of information and the use of neural network algorithms [3]. However, the effectiveness of these strategies in capturing complex patterns and relationships that are naturally present in physiological data is limited. Advancements in deep learning (DL) techniques, specifically deep convolutional neural networks (DCNNs) [4–6], offer a hopeful opportunity to significantly enhance the accuracy and efficiency of affective state mining. DCNNs have shown remarkable efficacy in several areas, including speech and picture recognition, due to their ability to independently learn hierarchical features from unprocessed input. The use of DCNNs on physiological inputs has significant promise in enhancing emotion identification within the field of affective computing, hence facilitating the attainment of more accurate and dependable outcomes [7].

Human Affective State Mining is a growing field of study in affective computing and human–computer interaction. It involves the use of a DCNN and various physiological signals such as EDA, EMG, ECG, and EEG. The objective of this study field is to create intelligent systems that can accurately perceive and understand human emotional states. This is achieved by analyzing physiological signals, which are regarded as reliable and impartial indicators of human emotional states. DCNNs, a type of DL model, have shown promising results in the field of emotion recognition from physiological inputs [8,9]. DCNNs are well-suited for this task since they can independently learn hierarchical representations from raw input data. Consequently, the need for manual extraction of features is avoided. To identify emotional states, diverse techniques are utilized, such as analyzing speech, observing psychophysiological responses, interpreting facial expressions, and examining social media posts. Physiological observations obtained using medical sensors allow for the continuous and real-time monitoring of emotional states.

Emotions are triggered as a result of a safety-critical incident. The response is elicited through the activation of the neurological system, assessment of cognitive, and subsequent behavioral reaction [10,11]. In order to experience an emotion, it is necessary for the body to maintain a state of physiological arousal, which depends on the individual's perception of the situation's importance. According to the theoretical framework proposed by Izard [12], there are four distinct systems that are responsible for the activation of emotions. The aforementioned systems encompass the neural system, which operates by administering electrical stimulation to the brain; the sensorimotor system, which elicits sensations through facial expression and body posture; the motivation system, which generates emotions derived from sensory experiences and drives; and the cognitive system, which is involved in the processes of categorization, evaluation, and comparison. EEG signals, which measure electrical activity in the brain, have been widely used in affective state mining [13]. Studies have shown that distinct patterns of brain activity, as measured by EEG signals, are linked to various emotional states [14]. ECG and EMG signals, which respectively assess the electrical activity in the heart and muscles, have also been utilized for the purpose of recognizing emotions [15]. Different emotional states have been linked to variations in heart rate variability and muscle tension. EDA signals, which measure changes in skin conductance, provide further information that can be used to derive affective states [16]. Given that emotional arousal has a known impact on EDA, it can be used as a reliable measure for identifying emotions.

Figure 6.1 shows the valence–arousal emotional model, which is a flat picture with four quadrants. Excitation and pleasure are affective states that have both high positivity and high energy. This puts them in the first part of the affective space.

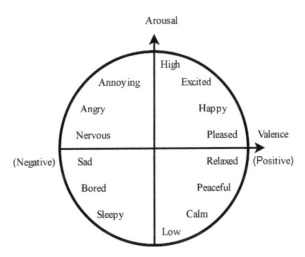

Figure 6.1   The valence–arousal circumplex of Russell's [15] model

Emotions with low valence and high arousal (LVHA) numbers, like anger and worry, are in the second quarter. In the third region are emotions like apathy, sadness, and despair. These emotions are different because they have low levels of valence and arousal, which are also known as LVLA states. The feelings of tranquility and happiness are in the fourth group. They are marked by high valence and low arousal (HVLA) levels. In terms of geometry, the idea of balance is right where the geometric centers of the axes meet.

Our chapter examines the categorization of emotions into five separate categories: the four quarters and neutral. Utilizing EEG data, this work makes two substantial contributions to the discipline of emotion analysis. To begin with, an innovative methodology is presented, which integrates refined convolutional neural networks (CNNs) with streamlined connectivity measures to discern and analyze emotions. Additionally, the research showcases the successful combination of multichannel EEG data with pre-trained deep-learning models, including Inception-v3, ResNet-50, and VGG-19. By utilizing the direct directed transfer function (DTF) method to investigate EEG connections, this study develops a flexible structure for analyzing emotions and highlights the capacity of DL for decoding complicated brain signals. The aforementioned contributions emphasize the distinctive methodology and effective implementation of sophisticated neural network architectures in the domain of emotional state evaluation through EEG data.

The report is organized in the following structure. Section 6.2 not only presents a summary of the preprocessing stages, methodology, and results but also evaluates the degree to which we were able to replicate the findings of the chapter. Section 6.3 provides detailed explanations of the datasets, preparation stages, model architectures, hyperparameters, and training procedures used. In Section 6.4 concise overview of the results and statistical analysis regarding the comparison of models and datasets is provided. Section 6.5 presents a comprehensive examination of the research findings concerning the stated objectives.

6.2   Related works

The research referenced in [17] investigates the application of DCNN in the analysis of EEG data for mining emotional states. The researchers employed a DCNN model to classify various emotional states by examining EEG data. The results suggest that the DCNN model demonstrated a significant level of accuracy in classifying emotions. This result highlights the favorable potential of utilizing EEG data for the task of extracting emotional states. Furthermore, a study [18] also utilizes DCNN to accurately identify emotions. The researchers have devised a unique design for DCNN known as CN-waterfall, which aims to identify physiological mood by including many forms of input. The model was trained and tested using a dataset comprising EEG and other physiological inputs. The results of the model in the field of emotion detection were deemed promising. The scholarly work reported in [19] provides a comprehensive overview of the area of emotion detection using EEG waves. The researchers analyzed various data mining techniques employed in the domain of EEG-based emotion recognition, including DCNN. The researchers highlighted the challenges and potential opportunities in this field. In their investigation, AlZoubi *et al.* [20] utilized EEG data obtained from the brain to determine the individual's emotional state. Employing a support vector machine (SVM) classifier to distinguish among 10 emotion classes results in a peak accuracy of 55%. The paper [21] provides a thorough description of the utilization of EEG signals for emotion identification. The authors discuss various data mining approaches, such as DL, that are used in EEG-based emotion identification. The difficulties and potential paths in this field are emphasized.

An alternative review paper [22] examines the use of ECG data in research on emotion recognition. The authors analyzed numerous research that utilized ECG signals and machine learning techniques, such as DCNN, to detect emotions. They discussed the healthcare uses of these approaches and the challenges in this field. Additional research conducted in [23] provides a comprehensive examination of emotion recognition through the utilization of EDA signals. The authors introduced several methodologies, including machine learning, that can be employed for emotion recognition based on exploratory data analysis. They highlighted the challenges and potential that exist in this field. A different system [24] was created to autonomously analyze emotions and accurately represent them using measurements from ECG, EDA, skin temperature, and an SVM classifier. Wei *et al.* [25] propose the Weight Fusion strategy as a novel classification technique to identify emotional states using various data such as ECG, EEG, respiration amplitude, and EDA. Pinto *et al.* [26] introduced a qualitative framework for examining emotions by utilizing physiological markers such as ECG, EDA, and EMG. Their investigation focused on determining if emotional state identification systems employ signals with higher informational capacity and identifying which signals are used.

Healey [27,28] produced one of the first databases about emotional physiology. A total of 24 participants were observed operating motor vehicles within the vicinity of Boston, and the collected dataset was supplemented with annotations indicating the respective levels of stress experienced by the drivers. Among the 24 replies collected from participants, a total of 17 are accessible to the public. The collected data comprises an ECG, galvanic skin response (GSR) recorded from the palms and feet, EMG extracted from the right trapezius muscle, and breathing patterns. The proposed method [29] introduces a succinct and autonomous representation of EEG responses for emotion identification. The suggested method differs from previous research by employing a two-part unsupervised generative model and segment-level feature extraction. The study [30] investigates the utilization of BCIs for the categorization and identification of emotions using EEG signals. The Dataset for Emotion Analysis using Physiological Signals (DEAP) dataset is utilized to demonstrate a technique for accurately delineating and categorizing various affective states.

There has been significant interest from several research areas in capturing the emotional state of users by observing physical manifestations, verbal intonation, and facial expressions. Lin *et al.* [31] present a technique for categorizing emotional states using a DCNN trained end-to-end. This approach is influenced by the progress made in the field of deep convolutional neural networks for image processing. The approach is evaluated using the DEAP database, which consists of EEG and peripheral physiological signal data. Another article [32] performs a meta-analysis of DL architectures used to retrieve biological data from different domains. This study investigates the application of DL architectures for pattern identification in sequences, signals, and image data. Another method [33] investigates the use of DL techniques, specifically DCNN, for detecting emotions based on physiological signals. The text refers to the AMIGOS dataset, which is used for classifying emotional states. This study compares the performance of DCNN and standard machine learning algorithms in classifying emotions. The MAHNOB-HCI database comprises two experimental studies [34]. Data from 30 subjects were collected, encompassing physiological signs, facial expressions, eye fixation, and EEG measurements. The original experiment consisted of 20 fervent movies selected from cinema and online sources. Participants observed photos and short videos illustrating human actions initially without a label, then subsequently with a visible label, as part of the second experiment, known as the Tag Agreement Experiment. The level of agreement among participants on the presented tag was evaluated for categories that were classified as either accurate or inaccurate. This research [35] further explores the use of CNN and deep neural networks (DNNs) for classifying emotions using EEG signals from the DEAP dataset. The DNN model achieved classification accuracies of 75.78% and 73.125% for valence and arousal, respectively. By comparison, the CNN model attained accuracy rates of 81.406% and 73.36% using the same parameters. These findings exhibited a higher level of performance in comparison to previous studies conducted on the DEAP dataset.

The study in [36] presents a multimodal dataset that is used to investigate human affective states. In a more specific manner, the study examines the EEG and peripheral physiological signals of a group of 32 volunteers who were exposed to music videos. This study presents a unique approach to stimulus selection, integrating video highlight recognition, and emotional markings obtained from the last. fm website, and an online evaluation. The primary focus of this work [37] centers around their methodology for measuring valence and arousal levels. This is achieved by combining face micro-expressions, EEG signals, GSR, and photoplethysmogram (PPG) signals. The model is evaluated using a subject-independent approach and the DEAP dataset. Furthermore, the authors present forecasts for the future use of facial micro-expressions and physiological data in the field of emotion recognition, while also acknowledging the constraints of the study. Several strategies have been suggested for emotion recognition using EEG signals, including time–frequency analysis, autoencoder with K-means clustering, essential EEG channel selection, domain adaptation methods, and angle space reconstruction [38].

CNNs are employed to discern individuals' emotional states by analyzing their EEG data. In 2019, Zeng *et al.* introduced a novel deep-learning network called SincNet-R, designed to enable the recognition of discrete emotional states [39]. In 2020, Shen *et al.* [40] proposed a novel approach for emotion recognition using

CNN and long short-term memory (LSTM). This method leverages the spatial, temporal, and frequency characteristics of multichannel EEG inputs. EEG signals are used for discerning emotions.

This study presents a unique methodology that employs dynamic directed transfer function (dDTF) and partial directed coherence (PDC) to convert a uni-dimensional EEG signal into a two-dimensional (2D) picture. This technique successfully represents the interconnections among brain cells. Once created, the 2D images are inputted into CNN architectures that have undergone training to identify five distinct emotional states. The experimental phase of the study, utilizing the MAHNOB-HCI and DEAP databases, demonstrates that ResNet-50 exhibits superior performance when applied to dDTF images within the alpha frequency range (8–13 Hz). The MAHNOB-HCI database achieved accuracy, F1-score, recall, and precision levels of 99.43%, 99.42%,99.44%, and 99.42%, respectively. Similarly, the DEAP database achieved scores of 98.17 for accuracy and 98.23 for F1-score. The recommended model, which utilizes multichannel EEG data and the effective link measure of dDTF in conjunction with ResNet-50, proves to be a useful tool for comprehensive brain function analysis.

6.3   Materials and methods

This section provides a detailed description of the experimental infrastructure used in this inquiry, as shown in Figure 6.2 through a block diagram that outlines the overall framework. The efficacy of the suggested methodology is assessed by using the MAHNOB-HCI and DEAP datasets. The MAHNOB-HCI database encompasses a diverse array of emotional stimuli, whereas the DEAP database focuses on physiological signals especially recorded during the observation of music videos. During preprocessing, EEG data undergoes cleaning and filtering to eliminate noise

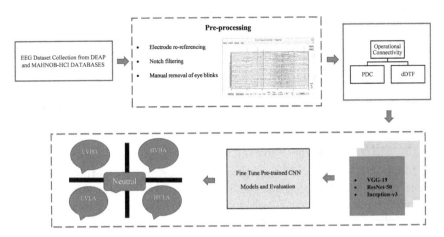

Figure 6.2   Illustration of the proposed emotion recognition system's block diagram

and artifacts that could potentially impact the identification of emotions. The dDTF technique reveals the connectivity patterns among the 32 EEG bands, generating 2D visualizations of brain information flow.

6.3.1   DEAP dataset description

The DEAP dataset, which is used for emotion analysis, was released in 2014. Furthermore, this dataset is not only one of the largest publicly available collections of data in the field of emotional computing, but it also includes a wide range of physiological and visual signals. The DEAP datasets are partitioned into two components:

1. A comprehensive compilation of 120 music videos, each lasting 1 min, assessed by 14–16 volunteers who evaluated them based on valence, arousal, and dominance.
2. A subset of 40 music videos was selected, each correlating to the EEG and physiological signals of one of the 32 subjects. The evaluation of each film was conducted using the same criteria of valence, arousal, and dominance as in the initial phase.

Only the EEG signal-containing second section of the DEAP dataset is used for this report. The Biosemi ActiveTwo device was used to gather four EEG signals. This device can record 32 EEG channels and has an adjustable sampling rate. The DEAP signals were initially obtained at a sampling rate of 512 Hz. However, the dataset's authors have also included a preprocessed EEG signal set at 128 Hz. This preprocessed set has undergone downsampling and includes frequency filters and other beneficial preprocessing techniques.

*   Each of the 32 participants is provided with access to the subsequent preprocessed information:
    The data consists of an array of 8064 elements. Each element represents a recording from one of the 40 channels and 40 music videos. Therefore, the array has a dimension of $40 \times 40 \times 8064$.
*   The video consists of 8064 recordings per channel, which is a result of the 63 s trial period (3 s pre-trial baseline + 60 s trial) and a sampling rate of 128 Hz ($63 \times 128 = 8064$).
*   Labels: The array consists of 40 rows and 4 columns, with each cell including annotated values for valence, arousal, dominance, and linkage of each of the 40 music videos (Figure 6.3).

6.3.2   MAHNOB dataset description

The MAHNOB dataset for affect recognition was released in 2012 [41–43]. This multimodal data collection encompasses not just eye gaze data but also audio, video, and physiological inputs. All input is synchronized and labeled based on the emotive qualities of valence and arousal. Four discrete types of investigations have been conducted:

*   In the first set of experiments, participants had to indicate their level of emotional positivity and intensity in response to a video stimulus while watching a video.

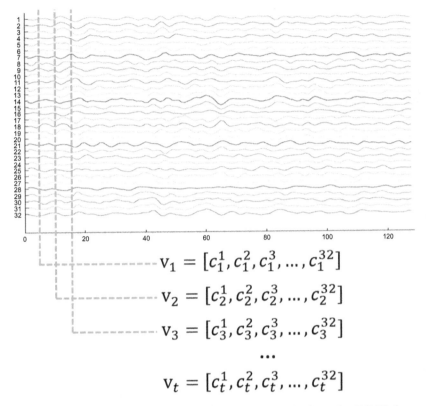

$$v_1 = [c_1^1, c_1^2, c_1^3, \ldots, c_1^{32}]$$

$$v_2 = [c_2^1, c_2^2, c_2^3, \ldots, c_2^{32}]$$

$$v_3 = [c_3^1, c_3^2, c_3^3, \ldots, c_3^{32}]$$

$$\ldots$$

$$v_t = [c_t^1, c_t^2, c_t^3, \ldots, c_t^{32}]$$

Figure 6.3   32-channel unprocessed EEG data example from the DEAP dataset

- In the other three experimental setups, a label was placed at the bottom of the screen; the connection between the label and the displayed film was uncertain. In this case, participants were asked to evaluate the significance of the title in connection to the video.

This study exclusively utilizes the data received from the initial category of studies. The EEG signals were acquired using the Biosemi ActiveTwo system, which is the same apparatus used for the collection of the DEAP dataset. Therefore, the EEG signals also include 32 channels. However, it is important to note that the MAHNOB signal was obtained at a frequency of 256 Hz, as opposed to the standard 512 Hz. Unlike DEAP, MAHNOB does not provide a preprocessed version of the dataset. Instead, it provides the raw, unprocessed data files that have been collected, which include EEG (.bdf) files (Figure 6.4).

6.3.3 Datasets preprocessing

The EEG data, consisting of 32 channels, were subjected to preprocessing using the EEGLAB toolbox inside the MATLAB program [36]. The datasets used in this study were acquired from the MAHNOB-HCI and DEAP databases. In the

Figure 6.4   MAHNOB-HCI dataset comprising three consecutive frames [41]

preprocessing stage of MAHNOB-HCI, the EEGs underwent re-referencing using the averaging approach. Following that, three finite impulse response filters were used on the EEG with cutoff frequencies of 0.5, 45, and 49–51 Hz. Furthermore, meticulous efforts were made to eliminate distortions caused by motion, blinking, and compression. In the study, a total of three patients were removed from the analysis owing to the significant presence of artifacts in their EEG signals. Consequently, further analysis and processing were conducted on the remaining 24 participants. All necessary procedures for the DEAP database have been executed, except for the average re-referencing stage.

6.3.3.1   Affective connectivity

Non-data-driven techniques, such as the continuous wavelet transform (CWT) and short-time Fourier transform (STFT), are often used for the conversion of 1D signals into 2D images. These transformations are well recognized for their ability to provide time–frequency representations. In contrast, model-based methodologies use techniques such as connectivity-based methods. All of these strategies are considered to be feasible alternatives. The main aim of this study was to assess the comparative efficacy of connectivity-based approaches against transformation-based methods in converting 1D data into pictures suitable for efficient emotion identification using the CNN model. During the execution of cognitive processes, the brain's network provides valuable insights using the notion of efficient connectivity. Mullen *et al.* [44] define effective connectivity as the presence of accidental or uneven interdependence between various brain areas. This methodology elucidates the transmission of information across distinct brain areas via EEG channels by selectively examining a certain frequency component. The objective is to transform unidimensional EEG data into a bidimensional depiction. The Granger-Causality approach, which is well-established for assessing effective connectivity, is computed in the frequency domain [45]. The validity of the produced model was later verified by evaluating the residuals' whiteness, consistency %, and stability, using the selected parameters.

6.3.3.2   dDTF measure

The dDTF is a method used to estimate effective connectivity in the brain by analyzing frequency domain data. It is based on the concept of conditional Granger-Causality [45]. Effective connection refers to the level of influence that a particular neurological system or region has over another, providing vital insights into the direction of information transmission within the brain. The DTF acts in the frequency domain, which is an expansion of the Granger-Causality framework. This allows for the analysis of the differences in how different areas of the brain interact with each other at different frequencies. The study employs the dDTF approach to evaluate EEG data to identify directional connection patterns among the 32 EEG bands. This leads to the creation of a 32×32 graphic that visually represents the flow of information between different brain areas across various frequencies. This innovative approach improves the accuracy of emotion detection models and adds depth to our understanding of brain connectivity by including frequency-specific data in the analysis.

6.3.3.3 PDC measure

PDC and dDTF are often used as estimators in the field of neuroscience to evaluate the transmission of information inside the brain [46]. The PDC is a quantitative measure that provides valuable information about the specific connections between different regions of the brain, therefore improving our understanding of how different parts of the brain interact with each other. PDC is a statistical metric that quantifies the degree of linear correlation between two signals while taking into consideration the influence of all other signals. It is based on the concept of partial coherence. PDC enhances the concept of brain connection by focusing on directed connectivity, allowing researchers to determine the extent and direction of information flow between different areas of the brain.

6.3.4 Pre-trained and convolutional neural network versions

CNNs are a specialized family of neural networks that find extensive use in the domain of biological signal processing and classification tasks [47–49]. The architecture of the network consists of many layers, including convolutional, pooling, batch normalization, fully connected (FC), and softmax. Convolutional layers are used to extract profound features at an elevated level. Subsequently, the pooling layers use maximum or average operators to decrease the dimensions of the feature maps, therefore removing the most crucial features. The primary purpose of the completely linked layers is to ease the transformation of the obtained data, hence enabling classification via the use of the softmax layer. The softmax function is a fundamental mathematical process often used in the last layer of a neural network. The algorithm computes the likelihood that the given input is a member of each of the existing classes. CNNs are used to enhance the accuracy of classification tasks, owing to their advantageous characteristics, including generalizability and flexibility. The present research employs three well-established pre-trained CNN models obtained from the ImageNet database. The use of models is applied to categorize EEG data into five distinct emotional states.

(i) VGG-19

VGGNet, commonly known as Visual Geometry Group, achieved the second position in the ILSVRC2014 competition [50]. The VGG-19 model has a capacity of 144 million parameters and is capable of processing color images with dimensions of 224 × 224 pixels (Figure 6.5). The concise configuration of the VGG-19 can be observed in Table 6.1.

(ii) **Inception-v3**

In the ILSVRC2015 competition, Inception version 3, often known as Inception-v3, achieved the second-place ranking [51]. The term "Inception" is derived from the inclusion of the Inception module, which has four parallel convolutional layers. This strategy significantly improves overall efficiency. The Inception-v3 model has a parameter count of 23.9 million, making it proficient in the effective processing of color images with dimensions of 299 × 299. Figure 6.6 depicts a concise representation of the structure of Inception-v3.

(iii) **ResNet-50**

The residual network (ResNet) [52] was awarded the ILSVRC2015 prize. The ResNet design integrates several identity shortcut connections to address the challenge of the vanishing gradient problem in CNNs. The ResNet architecture has many versions, each characterized by a different number of convolutional layers. These variations include models with 18, 34, 50, 101, and 152 convolutional layers. The performance of ResNet improves as the number of weighted layers increases. However, this also results in increased network complexity, computational load, and the requirement for more powerful hardware. The selection of ResNet-50 was based on its exceptional

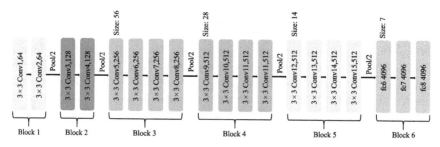

Figure 6.5   Block diagram of VGG-19

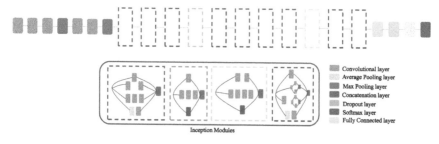

Figure 6.6   A compact block diagram illustrating Inception-v3

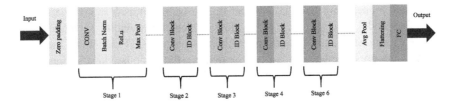

Figure 6.7   A block diagram of ResNet-50

performance and compatibility with hardware limitations. ResNet-50 is designed to handle color images with dimensions of 224 × 224 pixels and consists of a total of 25.6 million parameters. Figure 6.7 displays the condensed representation of the architecture of ResNet-50.

6.3.5   Evaluation criteria

The evaluation of the recognition system's efficacy was conducted using the 10-fold cross-validation approach. Three distinct variations of pre-trained CNNs were individually fine-tuned by the use of nine-fold cross-validation on DTF pictures. Following this, the previously mentioned models were assessed on an independent fold. The photographs were later rearranged in a random order, and this process was iterated a total of 10 times. The metrics of accuracy, precision, recall, and F1-score were then calculated [53]. Ultimately, the mean and standard deviation figures were given.

6.4   Results and discussion

The EEG signals from 32 channels obtained from the MAHNOB-HCI database (27 people) and DEAP database (32 participants) were preprocessed using MATLAB software (version 2019a). The completion of the installation of the EEGLAB toolbox (version 2021.0) has been achieved. In the end, the EEG data of three patients were removed from the analysis due to the significant presence of many artifacts. The remaining 24 patients received further processing. Considering the diverse variety of EEG signal durations included within this particular database, ranging from 34.9 to 117 s, our research focus will be directed toward examining signals that possess a minimum duration of 30 s. The MAHNOB and DEAP databases consist of a compilation of 10 different model orders. The MAHNOB-HCI database uses a window length of 5 s and a step size of 1 s. As a consequence, a signal was produced, leading to an output of 26 frames, each lasting for a length of 30 s. In addition, it has been determined that the window length and interval for the DEAP database are set at 6 and 1 s, respectively. As a result, the signal generated a total of 55 phases, each lasting for a length of 60 s. The estimation of DTF and PDC values for the MAHNOB-HCI and DEAP databases involves the use of 32 EEG channels per person. The estimate is conducted in five conventional frequency bands, namely delta, theta, alpha, beta, and gamma, between time intervals of 5 and 6 s. The estimation process utilizes the parameters of the developed multivariate auto-regressive model. The set of 32 × 32 values, which correspond to 32 channels, are acquired using efficient connectivity methods such as dDTF and PDC measures. These values are regarded as pictures with uniform dimensions. The aforementioned photos are then used as input for CNN models that have been pre-trained. The dimensions of the images produced by dDTF and PDC, namely 32 × 32 pixels, do not align with the input dimensions of the pre-trained models. The pixels in the dDTF and PDC images are enlarged. In contrast, the dimensions of each pixel in dDTF and PDC were augmented by a factor of almost seven. The conversion of 1D EEG data into visual representations is accomplished by the use of effective connection mea-surements. The MAHNOB-HCI and DEAP datasets jointly acquired a total of 12,480 dDTF images. This figure was computed by multiplying 26 window images by the number of subject shots and then by 20 video segments. Additionally, the datasets included 70,400 PDC images. This count was determined by multiplying 55 window images by the number of subject images.

The performance of the classifier was assessed on five unique emotional states using a nine-fold fine-tuning and one-fold restricting approach. Assessment was conducted using a range of metrics, including accuracy, precision, recall, and F1-score. Following the completion of 10 rounds of this procedure, the mean and standard deviation of each measurement were calculated and recorded. The selec-ted loss function was cross entropy, and the optimization phase used the ADAM method, which is known for its adaptive moment estimation capabilities. The learning rate was initially assigned a value of 0.0004, the squared gradient decay factor was assigned a value of 0.99, the minimum batch size was set to 32, and the maximum epoch count was set to 0.0001. A premature termination strategy was included, whereby the training process would be halted if the accuracy value (or loss function) remained constant at 100% (insignificant) for five consecutive epochs.

6.4.1 MAHNOB-HCI database results

Table 6.1 presents the outcomes obtained from the classification of the three pre-trained CNNs that underwent fine-tuning on PDC pictures. The classifica-tion was performed using the 10-fold cross-validation methodology, and the results include all frequency bands. The alpha frequency band exhibited higher levels of accuracy in comparison to other frequency bands when applied to PDC images, as seen by the clear results. The use of the ResNet-50 model resulted in achieving the highest values of 96.25%, 96.23%, 96.28%, and 96.23% for accuracy, precision, recall, and F1-score, respectively, within the alpha fre-quency range. Specifically, Inception-v3 attained an average accuracy of 95.85%, precision of 95.83%, recall of 95.87%, and an F1-score of 95.83%. Notably, the difference between these metrics was just 1%.

Table 6.2 displays the outcomes obtained from the classification of dDTF pictures using three pre-trained CNNs that underwent fine-tuning via the use of the 10-fold cross-validation methodology across all frequency bands.

Table 6.1   The classification results achieved by the proposed technique across all frequency bands are shown via the implementation of 10-fold cross-validation

Models	Bands	F1-score (%)	Precision (%)	Accuracy (%)	Recall (%)
VGG-19	Delta	85.20	85.22	85.25	85.26
	Theta	85.42	85.75	85.45	85.78
	Alpha	87.32	87.33	87.38	87.42
	Beta	86.02	86.03	86.05	86.10
	Gamma	85.72	85.72	85.75	85.78
ResNet-50	Delta	93.22	93.23	93.24	93.25
	Theta	93.61	93.63	93.64	93.66
	Alpha	96.23	96.23	96.26	96.28
	Beta	94.50	94.52	94.53	94.55
	Gamma	94.20	94.16	94.22	94.24
Inception-v3	Delta	92.07	92.08	92.10	92.14
	Theta	92.35	92.37	92.39	92.42
	Alpha	95.83	95.83	95.85	95.87
	Beta	93.26	93.25	93.29	93.34
	Gamma	93.05	93.05	93.07	93.12

Table 6.2   The proposed method achieved classification results across all frequency bands by employing a combination of pre-trained CNNs and dDTF images from the MAHNOB-HCI database, in combination with the cross-validation with the 10-fold technique

Models	Bands	F1-score (%)	Precision (%)	Recall (%)	Accuracy (%)
VGG-19	Delta	86.16	86.10	86.35	86.12
	Theta	86.39	86.42	86.75	86.45
	Alpha	88.30	88.29	88.30	88.30
	Beta	87.56	87.75	87.65	87.54
	Gamma	87.76	87.70	87.61	87.50
ResNet-50	Delta	97.31	97.35	97.20	97.30
	Theta	98.47	98.50	98.25	98.25
	Alpha	99.42	99.42	99.44	99.43
	Beta	98.54	98.67	98.42	98.44
	Gamma	98.29	98.55	98.37	98.35
Inception-v3	Delta	97.33	97.26	97.36	97.30
	Theta	97.35	97.27	97.36	97.30
	Alpha	98.55	98.55	98.55	98.55
	Beta	97.60	97.26	97.39	97.32
	Gamma	97.11	97.29	97.31	97.25

ResNet-50 demonstrated exceptional performance in accurately categorizing five emotional states from dDTF pictures, outperforming other models. The ResNet-50 model achieved superior performance in terms of accuracy, precision, recall, and F1-score within the alpha region. The precise figures observed were

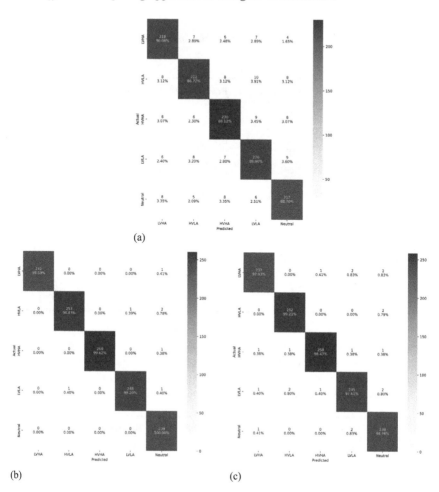

Figure 6.8   The confusion matrices of the proposed method utilizing various pre-trained CNNs (a) VGG-19, (b) ResNet-50, and (c) Inception-v3 and dDTF images in the alpha from the MAHNOB-HCI database

99.43%, 99.42%, 99.44%, and 99.42%, in that order. The Inception-v3 model demonstrated consistent accuracy ratings of 98.55% across many evaluations (Figure 6.8).

6.4.2   DEAP database results

The findings of categorizing PDC pictures across all frequency bands using three pre-trained CNNs that were tweaked using the 10-fold cross-validation approach are shown in Table 6.3. The alpha frequency band had the best accuracy among all models when applied to PDC pictures. The ResNet-50 model achieved the best levels of accuracy, precision, recall, and F1-score in the alpha frequency band, with

Table 6.3 Classification outcomes of the proposed method across all frequency bands utilizing various pre-trained CNNs and PDC images from the DEAP database and the 10-fold cross-validation technique

Models	Bands	F1-score (%)	Precision (%)	Accuracy (%)	Recall (%)
Inception-v3	Delta	91.21	91.24	91.25	91.27
	Theta	91.42	91.41	91.44	91.47
	Alpha	94.07	94.10	94.12	94.14
	Beta	92.54	92.52	92.56	92.59
	Gamma	92.20	92.18	92.21	92.23
ResNet-50	Delta	91.16	91.22	91.20	91.26
	Theta	91.50	91.55	91.52	91.54
	Alpha	95.69	95.70	95.73	95.77
	Beta	93.29	93.33	93.36	93.42
	Gamma	93.21	93.22	93.24	93.28
VGG-19	Delta	83.35	83.35	83.39	83.41
	Theta	83.70	83.75	83.76	83.78
	Alpha	85.38	85.41	85.42	85.45
	Beta	84.42	84.42	84.44	84.45
	Gamma	84.12	84.14	84.15	84.16

values of 95.73%, 95.70%, 95.77%, and 95.69% correspondingly. When comparing the outcomes seen in the PDC images from the MAHNOB-HCI database, it was found that Inception-v3 obtained the second highest ranking, exhibiting mean scores of 94.10%, 94.10%, 94.14%, and 94.07% for accuracy, precision, recall, and F1-score, respectively.

Among all models, ResNet-50 showed the highest accuracy in reliably recognizing five affective states from dDTF images, with the alpha frequency band doing the best. Inception-v3 and VGG-19 were outperformed by ResNet-50. The ResNet-50 model attained maximum accuracy, precision, recall, and F1-score of 98.16%, 98.16%, 98.16%, and 98.16%, respectively, while using the alpha frequency band. Figure 6.9 displays the confusion matrices for the several pre-trained models in the alpha region.

Significant success was achieved in this work via the use of DL and efficient connection approaches to automate the identification of emotional states. By implementing the ResNet-50 architecture and utilizing the dDTF technique inside the alpha frequency range, our analysis of pictures yielded remarkable accuracy rates of 99.43% and 98.16%. The analysis of EEG signals from 32 channels gathered from the MAHNOB-HCI and DEAP databases was conducted. A significant contribution of this research is in the innovation of a methodology for generating 2D visual representations from 32-channel 1D EEG recordings. The previously mentioned objective may be accomplished by feeding the data into a pre-existing CNN structure, which has been trained in advance, using parameters that measure the effective connectivity of the brain. Various methods, such as the CWT and STFT, may use classical time–frequency data to convert 1D signals into 2D visuals.

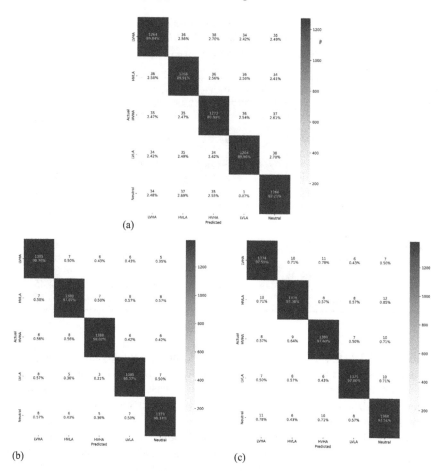

Figure 6.9   The confusion matrices generated by the proposed method when various pre-trained CNNs (VGG-19, ResNet-50, and Inception-v3) are applied to dDTF images from the database of DEAP in the bands of frequency of alpha, respectively

In contrast, we have used the dDTF and PDC metrics, which are established measures of effective connectivity.

According to Tables 6.1 and 6.2, it can be seen that the alpha frequency band exhibited the best level of accuracy when used in the fine-tuned CNNs to recognize the five emotional states as mentioned in the MAHNOB-HCI database. Following that, the beta, gamma, theta, and delta frequency bands occurred. In addition, it was observed that the alpha frequency band exhibited the best degree of accuracy when compared to all other fine-tuned CNNs used in the analysis of the DEAP database. The data reported in Tables 6.3 and 6.4 demonstrate this fact. Following that, the series of frequency bands included the gamma, beta, theta, and delta bands.

Table 6.4   Classification outcomes of the proposed method across all frequency bands utilizing various pre-trained CNNs and dDTF images from the DEAP database in conjunction with the 10-fold cross-validation technique

Models	Bands	F1-score (%)	Precision (%)	Accuracy (%)	Recall (%)
Inception-v3	Delta	97.14	97.35	97.20	96.50
	Theta	97.35	97.42	97.35	97.05
	Alpha	97.53	97.53	97.54	97.53
	Beta	97.17	97.30	97.25	96.86
	Gamma	97.28	97.57	97.50	96.73
ResNet-50	Delta	97.10	97.12	96.85	97.05
	Theta	97.05	97.10	96.83	96.92
	Alpha	98.16	98.16	98.16	98.16
	Beta	97.13	97.16	96.95	97.04
	Gamma	97.15	97.20	96.91	97.01
VGG-19	Delta	87.62	87.55	87.05	87.34
	Theta	88.50	88.47	88.10	88.25
	Alpha	89.95	89.95	89.95	89.95
	Beta	89.50	89.47	89.08	89.34
	Gamma	88.25	88.20	87.86	88.10

Therefore, while examining the two databases, it is evident that the alpha frequency band is more effective in distinguishing the five specified affective states compared to other frequency bands. Concerning the transmission of information across different parts of the brain, we discovered that the alpha band demonstrates superior discriminative behavior in comparison to other frequency bands. This finding was both informative and intriguing. Regarding alpha-effective connection using dDTF and PDC, our results align with previous studies investigating different connectivity properties. During music listening, the connection of dDTF is more pronounced in the frontal and parietal areas for the alpha, beta, and gamma frequency bands, compared to other frequency bands.

After analyzing Tables 6.1–6.4, we concluded that dDTF images are more effective and appropriate than PDC images in studying and identifying emotional states. The accuracy values for dDTF images were 99.43% for the MAHNOB-HCI database and 98.16% for the DEAP database, while the accuracy values for PDC images were 96.26% and 95.73% for the respective databases. The models ResNet-50 and Inception-v3 had the highest recognition scores for five different emotional states in the two datasets, as indicated in Tables 6.1–6.4. ResNet-50 and Inception-v3 utilize residual and Inception modules, respectively, which consist of convolutional layers structured in a specific and intricate manner. The residual module bypasses the input-to-output paths, while the Inception module utilizes parallel convolutional layers. Furthermore, these networks employ batch normalization to expedite convergence and mitigate overfitting.

Table 6.5 provides a comparative analysis of the outcomes of this inquiry and the most recent and pertinent research endeavors that delineate emotional states via the utilization of EEG signals sourced from identical databases. The valence–

Table 6.5   A comparison of studies utilizing EEG data to identify emotions is provided

Author	Databases	Method	Accuracy (%)
Yang *et al.* [54]	DEAP	Sample entropy, empirical mode decomposition, and support vector machine	93.2%
Zheng *et al.* [55]	DEAP	Regularized graph differential entropy extreme learning machine	69.67%
Soroush *et al.* [56]	DEAP	Multilayer perceptron, Poincare plane, multiclass support vector machine, and K-nearest neighbors	89.76%
Li *et al.* [57]	DEAP, MAHNOB-HCI	Phase locking value, SVM, generalized extreme learning machine	68% in MAHNOB-HCI, 62% in DEAP
Dai *et al.* [40]	DEAP	Fourth-order convolutional recurrent neural network with differential entropy	94.22% (valence), 94.58% (arousal)
Suggested method	MAHNOB-HCI, DEAP	The DL models often referred to as DTF, PDC, ResNet-50, Inception-v3, and VGG-19 are well-recognized and used in academic and industrial research	98.16% (DEAP, dDTF), 99.43% (dDTF, MAHNOB-HCI)

arousal paradigm was used to analyze emotional states in each of these trials. A significant number of these tests used a classification system that categorized emotional states into binary groups, using the concepts of valence and arousal. In a more specific manner, the researchers conducted a comparison of situations characterized by low physiological activity and those characterized by high physiological activation. Additionally, they examined states associated with good emotional worth and states associated with negative emotional value. On the contrary, to enhance precision, we undertook a comprehensive examination of emotional states. Five separate classifications were found, as described in the emotional states section. The classifications included in this categorization are as follows: positive valence and high arousal, positive valence and low arousal, negative valence and either low arousal or high arousal, and near zero valence and arousal, which corresponds to the neutral class. The research demonstrates a higher level of accuracy compared to previous tests that used traditional machine-learning techniques and alternative DL approaches. This finding provides strong evidence in favor of the suggested strategy. Therefore, this work distinguishes itself from other studies of a similar kind by using the dDTF effective connectivity picture obtained from EEG data and employing robust fine-tuned DL CNNs, resulting in improved accuracy. Based on the results shown in Table 6.5, our study has demonstrated significant achievements in the automation of emotional state recognition across two datasets, resulting in the most positive outcomes seen so far. As a result, our proposed approach aims to use the enhanced technique on additional classifications of EEG data.

6.5 Conclusion

A comprehensive examination was carried out to use the multichannel EEG signals to discern five emotional states while subjects viewed video clips that evoked strong emotional responses and video clips from music. The researchers used the MAHNOB-HCI and DEAP databases in their study. The research included the use of two reliable connection metrics, namely dDTF and PDC, in conjunction with many well-established pre-trained DL algorithms. The utilization of the ResNet-50 architecture in conjunction with the dDTF approach specifically applied to images within the alpha frequency band, yielded remarkably high accuracy rates of 99.43% and 98.16% in the classification of five distinct emotional states. These results were obtained from the analysis of data sourced from the MAHNOB-HCI and DEAP databases, respectively. The aforementioned findings have been obtained via the process of categorizing the images. The model presented in this study, which is based on the obtained data, demonstrates a higher level of effectiveness in the analysis of brain activity compared to earlier studies conducted in recent years. This is drawn from the information that has been uncovered. The suggested technique using dDTF and PDC with ResNet-50 for emotion categorization from EEG data shows promise; however, there are drawbacks. Specific datasets may limit generalizability, requiring cross-dataset validation for varied demographics and emotional settings. Pre-trained models imply universal applicability, encouraging fine-tuning or training from scratch for flexibility. Dataset emotion identification accuracy is critical, highlighting the necessity for uniform annotation techniques. The study's narrow emphasis on emotions necessitates more research into a wider spectrum. Future studies should include real-time implementation, hybrid techniques incorporating numerous physiological inputs, and model interpretability. A longitudinal study of the methodology's stability would improve its practicality in comprehending dynamic emotional states. Addressing these constraints and exploring these future research options will improve and broaden the approach's EEG-based emotional state analysis applications.

References

[1] M. N. I. Shuzan, M. H. Chowdhury, M. S. Hossain, M. S. A. Hossain, and M. E. H. Chowdhury, "Machine-learning-based emotion recognition in arousal–valence space using photoplethysmogram signals," in *Affective Computing in Healthcare: Applications based on Biosignals and Artificial Intelligence*, Bristol: IOP Publishing, 2023, pp. 1–5.

[2] P. J. Bota, C. Wang, A. L. N. Fred, and H. P. Da Silva, "A review, current challenges, and future possibilities on emotion recognition using machine learning and physiological signals," *IEEE Access*, vol. 7, pp. 140990–141020, 2019.

[3] M. S. I. Sumon, T. Ahmed, M. M. Khan, M. M. A. Mahedi, M. N. I. Shuzan, and M. J. Al Nahian, "Detection of diabetes using 1D convolutional neural

network from short time PPG signal," in *2023 International Conference on Information and Communication Technology for Sustainable Development (ICICT4SD)*, 2023, pp. 372–376, doi:10.1109/ICICT4SD59951.2023. 10303575.

[4] T. Rahman, A. Akinbi, M. E. H. Chowdhury, *et al.*, "COV-ECGNETA: COVID-19 detection using ECG trace images with deep convolutional neural network," *Health Information Sciences and Systems*, vol. 10, no. 1, p. 1, 2022.

[5] R. Yuvaraj, A. Baranwal, A. A. Prince, M. Murugappan, and J. S. Mohammed, "Emotion recognition from spatio-temporal representation of EEG signals via 3D-CNN with ensemble learning techniques", *Brain Sciences*, vol. 13, p. 685, 2023, doi:10.3390/brainsci1304068.

[6] A. Hassouneh, A. M. Mutawa, and M. Murugappan, "Development of a real-time emotion recognition system using facial expressions and EEG based on machine learning and deep neural network methods," *Informatics in Medicine*, vol. 20, p. 100372, 2020.

[7] F. Patlar Akbulut, "Hybrid deep convolutional model-based emotion recognition using multiple physiological signals," *Computer Methods in Biomechanics and Biomedical Engineering*, vol. 25, no. 15, pp. 1678–1690, 2022.

[8] H. Zhang, R. Gou, J. Shang, F. Shen, Y. Wu, and G. Dai, "Pre-trained deep convolution neural network model with attention for speech emotion recognition," *Frontiers in Physiology*, vol. 12, p. 643202, 2021.

[9] T. Puri, M. Soni, G. Dhiman, O. Ibrahim Khalaf, and I. Raza Khan, "Detection of emotion of speech for RAVDESS audio using hybrid convolution neural network," *Journal of Healthcare Engineering*, vol. 2022, 2022.

[10] R. S. Lazarus and C. A. Smith, "Knowledge and appraisal in the cognition—emotion relationship," *Cognition Emotion*, vol. 2, no. 4, pp. 281–300, 1988, doi:10.1080/02699938808412701.

[11] A. T. Beck and D. A. Clark, "An information processing model of anxiety: automatic and strategic processes," *Behaviour Research and Therapy*, vol. 35, no. 1, pp. 49–58, 1997.

[12] E. Palomba, "How to make learning a pleasure? Reflections on emotion, cognition, and educational strategies," in *EDULEARN23 Proceedings, IATED*, 2023, pp. 1051–1059.

[13] A. Appriou, A. Cichocki, and F. Lotte, "Modern machine-learning algorithms: for classifying cognitive and affective states from electro-encephalography signals," *IEEE Systems, Man, and Cybernetics Magazine*, vol. 6, no. 3, pp. 29–38, 2020.

[14] M. M. Rahman, A. K. Sarkar, M. A. Hossain, *et al.*, "Recognition of human emotions using EEG signals: A review," *Computers in Biology and Medicine*, vol. 136, p. 104696, 2021.

[15]   M. Egger, M. Ley, and S. Hanke, "Emotion recognition from physiological signal analysis: A review," *Electronic Notes in Theoretical Computer Science*, vol. 343, pp. 35–55, 2019.

[16]   J. Shukla, M. Barreda-Angeles, J. Oliver, G. C. Nandi, and D. Puig, "Feature extraction and selection for emotion recognition from electrodermal activity," *IEEE Transactions on Affective Computing*, vol. 12, no. 4, pp. 857–869, 2019.

[17]   M. G. R. Alam, S. F. Abedin, S. I. Moon, A. Talukder, and C. S. Hong, "Healthcare IoT-based affective state mining using a deep convolutional neural network," *IEEE Access*, vol. 7, pp. 75189–75202, 2019, doi:10.1109/ACCESS.2019.2919995.

[18]   N. Fouladgar, M. Alirezaie, and K. Främling, "CN-waterfall: A deep convolutional neural network for multimodal physiological affect detection," *Neural Computing and Applications*, vol. 34, no. 3, pp. 2157–2176, 2022, doi:10.1007/s00521-021-06516-3.

[19]   F. Mendoza-Palechor, M. L. Menezes, A. Sant'Anna, M. Ortiz-Barrios, A. Samara, and L. Galway, "Affective recognition from EEG signals: an integrated data-mining approach," *Journal of Ambient Intelligence and Humanized Computing*, vol. 10, no. 10, pp. 3955–3974, 2019, doi:10.1007/s12652-018-1065-z.

[20]   O. AlZoubi, R. A. Calvo, and R. H. Stevens, "Classification of EEG for affect recognition: An adaptive approach," in *AI 2009: Advances in Artificial Intelligence: 22nd Australasian Joint Conference*, Melbourne, Australia, December 1–4, 2009. Proceedings 22, Springer, 2009, pp. 52–61.

[21]   J. Liu, H. Wu, L. Zhang, and Y. Zhao, "Spatial-temporal transformers for EEG emotion recognition," October 2021, [Online]. Available at: http://arxiv.org/abs/2110.06553

[22]   J. Chennouf and R. Chiheb, "What machine learning (ML) can bring to the electrocardiogram (ECG) signal: A review," in X.-S. Yang, S. Sherratt, N. Dey, and A. Joshi (eds.), *Proceedings of Seventh International Congress on Information and Communication Technology*, Singapore: Springer Nature, 2023, pp. 61–69.

[23]   P. Romaniszyn-Kania, A. Pollak, M. D. Bugdol, *et al.*, "Affective state during physiotherapy and its analysis using machine learning methods," *Sensors*, vol. 21, no. 14, 2021, doi:10.3390/s21144853.

[24]   K. H. Kim, S. W. Bang, and S. R. Kim, "Emotion recognition system using short-term monitoring of physiological signals," *Medical and Biological Engineering and Computing*, vol. 42, pp. 419–427, 2004.

[25]   W. Wei, Q. Jia, Y. Feng, and G. Chen, "Emotion recognition based on weighted fusion strategy of multichannel physiological signals," *Computational Intelligence and Neuroscience*, vol. 2018, 2018.

[26]   G. Pinto, J. M. Carvalho, F. Barros, S. C. Soares, A. J. Pinho, and S. Brás, "Multimodal emotion evaluation: A physiological model for cost-effective emotion classification," *Sensors*, vol. 20, no. 12, p. 3510, 2020.

[27] J. A. Healey and R. W. Picard, "Detecting stress during real-world driving tasks using physiological sensors," *IEEE Transactions on Intelligent Transportation Systems*, vol. 6, no. 2, pp. 156–166, 2005.

[28] J. A. Healey, *Wearable and Automotive Systems for Affect Recognition from Physiology*, Massachusetts Institute of Technology, 2000.

[29] X. Zhuang, R. Rozgić, and M. Crystal, Compact Unsupervised EEG Response Representation for Emotion Recognition, 2014, doi: 10.0/Linux-x86_64.

[30] G. Placidi, P. Di Giamberardino, A. Petracca, M. Spezialetti, and D. Iacoviello, "Classification of emotional signals from the DEAP dataset," in *NEUROTECHNIX 2016 – Proceedings of the 4th International Congress on Neurotechnology, Electronics and Informatics*, SciTePress, 2016, pp. 15–21, doi:10.5220/0006043400150021.

[31] W. Lin, C. Li, and S. Sun, "Deep convolutional neural network for emotion recognition using EEG and peripheral physiological signal," in *Lecture Notes in Computer Science (including subseries Lecture Notes in Artificial Intelligence and Lecture Notes in Bioinformatics)*, Springer, 2017, pp. 385–394, doi:10.1007/978-3-319-71589-6_33.

[32] M. Mahmud, M. S. Kaiser, T. M. McGinnity, and A. Hussain, "Deep learning in mining biological data," *Cognitive Computation*, vol. 13, no. 1, pp. 1–33, 2021, doi:10.1007/s12559-020-09773-x.

[33] L. Santamaria-Granados, M. Munoz-Organero, G. Ramirez-Gonzalez, E. Abdulhay, and N. Arunkumar, "Using deep convolutional neural network for emotion detection on a physiological signals dataset (AMIGOS)," *IEEE Access*, vol. 7, pp. 57–67, 2019, doi:10.1109/ACCESS.2018.2883213.

[34] M. Soleymani, J. Lichtenauer, T. Pun, and M. Pantic, "A multi-modal affective database for affect recognition and implicit tagging," *IEEE Transactions on Affective Computing*, vol. 3, p. 1, 2011.

[35] S. Tripathi, S. Acharya, R. Dev Sharma, S. Mittal, and S. Bhattacharya, Using Deep and Convolutional Neural Networks for Accurate Emotion Classification on DEAP Dataset [Online]. Available at: www.aaai.org.

[36] S. Koelstra, C. Muh, M. Soleymani, *et al.*, "DEAP: A database for emotion analysis using physiological signals," *IEEE Transactions on Affective Computing*, vol. 3, no. 1, pp. 18–31, 2012, doi: 10.1109/T-AFFC.2011.15.

[37] N. Saffaryazdi, S. T. Wasim, K. Dileep, *et al.*, "Using facial micro-expressions in combination with EEG and physiological signals for emotion recognition," *Frontiers in Psychology*, vol. 13, 2022, doi:10.3389/fpsyg.2022.864047.

[38] J. Chen, T. Ro, and Z. Zhu, "Emotion recognition with audio, video, EEG, and EMG: A dataset and baseline approaches," *IEEE Access*, vol. 10, pp. 13229–13242, 2022, doi:10.1109/ACCESS.2022.3146729.

[39] H. Zeng, Z. Wu, J. Zhang, *et al.*, "EEG emotion classification using an improved SincNet-based deep learning model," *Brain Sciences*, vol. 9, no. 11, p. 326, 2019.

[40] F. Shen, G. Dai, G. Lin, J. Zhang, W. Kong, and H. Zeng, "EEG-based emotion recognition using 4D convolutional recurrent neural network," *Cognitive Neurodynamics*, vol. 14, pp. 815–828, 2020.

[41] S. Petridis, B. Martinez, and M. Pantic, "The MAHNOB laughter database," *Image Vision Computation*, vol. 31, no. 2, pp. 186–202, 2013.

[42] M. B. H. Wiem and Z. Lachiri, "Emotion classification in arousal valence model using MAHNOB-HCI database," *International Journal of Advanced Computer Science and Applications*, vol. 8, no. 3, 2017.

[43] S. Bilakhia, S. Petridis, A. Nijholt, and M. Pantic, "The MAHNOB mimicry database: A database of naturalistic human interactions," *Pattern Recognition Letters*, vol. 66, pp. 52–61, 2015.

[44] T. Mullen, "Source information flow toolbox (SIFT)," *Swartz Center for Computational Neuroscience*, pp. 1–69, 2010.

[45] H. Shahabi and S. Moghimi, "Toward automatic detection of brain responses to emotional music through analysis of EEG effective connectivity," *Computational Human Behavior*, vol. 58, pp. 231–239, 2016.

[46] T. F. Tafreshi, M. R. Daliri, and M. Ghodousi, "Functional and effective connectivity based features of EEG signals for object recognition," *Cognitive Neurodynamics*, vol. 13, pp. 555–566, 2019.

[47] A. Craik, Y. He, and J. L. Contreras-Vidal, "Deep learning for electro-encephalogram (EEG) classification tasks: A review," *Journal of Neural Engineering*, vol. 16, no. 3, p. 031001, 2019.

[48] O. Faust, Y. Hagiwara, T. J. Hong, O. S. Lih, and U. R. Acharya, "Deep learning for healthcare applications based on physiological signals: A review," *Computer Methods and Programs in Biomedicine*, vol. 161, pp. 1–13, 2018.

[49] A. Saeedi, M. Saeedi, A. Maghsoudi, and A. Shalbaf, "Major depressive disorder diagnosis based on effective connectivity in EEG signals: A convolutional neural network and long short-term memory approach," *Cognitive Neurodynamics*, vol. 15, pp. 239–252, 2021.

[50] K. Simonyan and A. Zisserman, "Very deep convolutional networks for large-scale image recognition," arXiv preprint arXiv:1409.1556, 2014.

[51] C. Szegedy, V. Vanhoucke, S. Ioffe, J. Shlens, and Z. Wojna, "Rethinking the inception architecture for computer vision," in *Proceedings of the IEEE Conference on Computer Vision and Pattern Recognition*, 2016, pp. 2818–2826.

[52] K. He, X. Zhang, S. Ren, and J. Sun, "Deep residual learning for image recognition," in *Proceedings of the IEEE Conference on Computer Vision and Pattern Recognition*, 2016, pp. 770–778.

[53] M. Sokolova and G. Lapalme, "A systematic analysis of performance measures for classification tasks," *Information Processing and Management*, vol. 45, no. 4, pp. 427–437, 2009.

[54] Y.-X. Yang, Z.-K. Gao, X.-M. Wang, *et al.*, "A recurrence quantification analysis-based channel-frequency convolutional neural network for emotion

recognition from EEG," *Chaos: An Interdisciplinary Journal of Nonlinear Science*, vol. 28, no. 8, p. 085724, 2018.

[55] J. Zhang, S. Zhao, W. Huang, and S. Hu, "Brain effective connectivity analysis from EEG for positive and negative emotion," in *Neural Information Processing: 24th International Conference, ICONIP 2017*, Guangzhou, China, November 14–18, 2017, Proceedings, Part IV 24, Springer, 2017, pp. 851–857.

[56] M. Z. Soroush, K. Maghooli, S. K. Setarehdan, and A. M. Nasrabadi, "Emotion recognition using EEG phase space dynamics and Poincare intersections," *Biomedical Signal Processing and Control*, vol. 59, p. 101918, 2020.

[57] P. Li, H. Liu, Y. Si, *et al.*, "EEG based emotion recognition by combining functional connectivity network and local activations," *IEEE Transactions on Biomedical Engineering*, vol. 66, no. 10, pp. 2869–2881, 2019.

Chapter 7

Deep neural network-based stress detection using biosignals

R. Bharathi Vidhya[1], S. Jerritta[1] and T. Thiyagasundaram[2]

Recently, electrocardiogram (ECG) signals have been used in affective computing to recognize and interpret human emotions, loneliness, pain, and other psychological states. In healthcare, human–computer interaction, and good health, stress detection is one application of affective computing that has been gaining attention. As one of the major factors in diagnosing stress, artificial intelligence uses sophisticated algorithms to examine physiological data that is difficult to interpret and find subtle signs of stress. Our article proposes a stress prediction model based on long short-term memory (LSTM). In this model, the temporal characteristics of the data are processed using recurrent neural networks, which are compatible with time-series data such as ECG signals. First, the ECG signals are normalized, and then the input is classified further to assess the performance of the proposed model. The features are extracted using a four-layered LSTM network and wavelet transform. Afterwards, the features are optimized to improve classification accuracy. Due to its high sensitivity and accuracy, the proposed model outperforms other methods of feature extraction such as kernel methods, principal component analysis, and embedders.

Keywords: Long short-term memory (LSTM); Electrocardiogram (ECG); Discrete wavelet transform (DWT); Stress detection; Affective computing

7.1 Introduction

Researchers have conducted numerous studies on identifying emotions from physiological signals in healthy individuals over the past few decades, either with a single physiological signal or a mixture of physiological signals [1]. GOQii conducted a Stress and Mental Health Study among 10K+ Indians as part of India Fit

[1]Department of ECE, Vels Institute of Science, Technology and Advanced Studies, India
[2]Department of NoiNadal, Sri Sairam Siddha Medical College and Research Centre, West Tambaram, India

2022–23, which concluded that 24% of Indians are experiencing stress in various forms [2,3]. Statistics and surveys by the American Institute of Stress indicate that 35% of people in 143 countries are stressed. The use of physiological signals to detect human emotions is widely considered to be a valuable tool [4]. Since stress is an individualized and complex phenomenon, there is no single solution that works for everyone [5]. It may be possible to improve the precision and dependability of stress detection systems by incorporating different modalities and taking individual variances into account [6]. A patient's emotional and mental state can be assessed by analyzing speech patterns, facial expressions, and physiological data [7–9]. As a result, healthcare providers can tailor their care plans based on this information and gain insights into a patient's mental health. In addition to coronary diseases and emotional well-being, acute stress has been linked to natural resistance suppression and cardiovascular disease [10,11]. The early detection and intervention of his/her emotional state can prevent the development of these health issues [12,13].

As a term, affective computing refers to the development of systems and devices that can detect stress. Affective computing involves several areas, such as the creation of affective states, and interpreting or identifying affective states [14]. Healthcare professionals can benefit from this emerging field in various ways, including enhancing patient care and improving the patient experience. The use of affective computing can be used to monitor the emotional states of patients based on their facial expressions, voice analysis, and physiological signals [15]. Mental health conditions can be detected early and monitored through the use of affective computing technologies. Patient engagement can be improved by personalizing healthcare interactions through affective computing. Empathetic virtual assistants and chatbots can respond to patients' needs empathically, enhancing their overall health status.

By analyzing facial expressions, body language, and other physiological signals, affective computing can assist in assessing and managing pain [14]. By utilizing this information, healthcare providers can tailor pain management strategies to the needs of each patient. As the successful application of affective computing depends on a deep understanding of human emotional states, affect recognition is of paramount importance. Additionally, these methods must be developed and implemented with care in order to take into account privacy issues, user consent, and ethical concerns. The types of data and application environment influence the type of machine learning (ML) technique used for stress detection [16]. Stress can be identified with the use of strong classifiers called support vector machines (SVMs), especially when working with structured data [3]. SVMs, however, may exhibit majority class bias when they are applied to imbalanced datasets containing unevenly distributed examples of stressed and non-stressed states. The result may be less than optimal performance in terms of stress detection, particularly if the dataset is unbalanced [17].

In cases where the correlation between features and stress is clear-cut, decision trees can be interpreted and are useful for detecting stress [18]. Decision trees are simple models, so they might not be able to identify deep relationships and patterns in the data, especially when identifying stress requires a complex interaction between several characteristics. In situations where local patterns are important,

K-nearest neighbor (KNN) can also be used to detect stress [7]. Recurrent neural networks that work well with sequential data include the long short-term memory (LSTM) model. Because LSTMs can capture temporal dependencies, they have been used to analyze time-series data in stress detection settings, including physiological signals and speech patterns. Because of their many benefits, LSTM models are a good choice for some stress detection applications, especially when working with sequential or time-series data.

7.2   Related works

A physiological signal can be used to detect stress by monitoring and analyzing various bodily responses associated with it. Stress levels can be determined by monitoring heart rate using sensors. Stress is often associated with an increased heart rate. Electrocardiogram (ECG) signals were used to derive heart rate variability, an indicator of one's ability to identify emotions in others [12]. It has been shown that changes in skin conductance can indicate emotional arousal, including stress [19]. Those who suffer from hypertension will have elevated switched capacitor levels, which are stable measures of stress [20]. Brainwave patterns can also be obtained using electroencephalogram (EEG) signals to detect stress-related brain activity. Multi-physiological signals could be used to provide a deeper understanding of the stress response of an individual [21]. Various muscle electromyogram (EMG) signals are also analyzed for the detection of multi-level stress [22].

It is possible to analyze and interpret complex patterns in physiological data using ML techniques. Analysis of facial expressions for signs of stress, such as furrowed brows, tense muscles, or frowning, uses facial recognition technology. The ability to recognize subtle facial cues associated with stress may be trained by using deep neural networks. A three-dimensional (3D) convolution algorithm is used to examine facial expressions in successive frames and record the spatio-temporal aspects [23]. Stress can be detected by tracking changes in physical activity, sleep patterns, or social interactions. It has been established that brain function is crucial for emotional regulation, which in turn indirectly influences sleep quality [24]. It identifies stress-induced changes in typing speed and errors by analyzing typing patterns. It is also possible to measure stress levels continuously using wearable devices like smartwatches and fitness trackers [25].

Physiological and emotional data must be handled securely and in compliance with privacy regulations while using these techniques [26]. We must obtain user consent during stress monitoring and be transparent about collected data use. The use of affective computing is subject to ethical considerations, including potential misuse of emotional data, user consent, and algorithmic bias. Transparency and ethical guidelines are critical to the development and deployment of these systems.

7.2.1   *Challenges and gaps identified in affective computing*

The use of affective computing for stress detection holds great promise, but there are several challenges to overcome if the technology is to be effective, ethical, and widely adopted. Stress is a subjective experience, and individuals may display different physiological and behavioral responses to it. It is challenging to develop a universal model that accommodates individual differences. Multi-modal data fusion, model complexity, and the requirement of diverse datasets (e.g., physiological signals, facial expressions, speech) pose challenging challenges for feature fusion, model complexity, and data diversity [19].

The collection of sensitive personal data is often a part of affective computing. It is critical to ensure robust data privacy, secure data storage, and ethical data usage to maintain user trust and comply with privacy regulations. There are many challenges to implementing stress detection systems in real-world settings, such as dealing with environmental noise, device compatibility, and user acceptance. There may be significant differences between real-world conditions and controlled research environments. It is crucial for gaining user trust and acceptance to ensure that affective computing models are interpretable and provide explanations for their predictions. Deep learning models are often perceived as black boxes.

The subjective nature of stress makes it challenging to establish standardized benchmarks and validation metrics for stress detection. The performance of different models needs to be compared using consistent evaluation methods. Monitoring chronic stress patterns over time is crucial to understanding chronic stress patterns. However, this raises concerns about user compliance, data storage, and habituation to monitoring devices. It can be challenging to integrate affective computing systems with existing healthcare, work, or educational systems due to seamless compatibility and interoperability.

7.2.2   Feasible proposed solution

A successful implementation of artificial intelligence (AI) can help detect stress effectively. Stress detection is primarily an ML process that involves data collection, preprocessing, and feature extraction. Feature engineering approaches are used to extract relevant data or generate features that can record stress-related patterns. To identify physiological or behavioral signs of stress, time-series analysis, frequency domain analysis, or statistical measure extraction may be used. Performance is improved by either engineering or removing pertinent features. In addition to ML, we have considered some popular AI techniques, such as LSTM, convolutional neural networks (CNNs), analysis of variance (ANOVA), KNN, and natural language processing (NLP) based modeling [1].

In CNNs, coupled convolution architectures with dropout mechanisms are used. It consists of a dense layer, a convolutional layer, an input layer, and an output layer. Data from the input is used to extract features from CNN. The overall performance of a CNN is significantly impacted by uneven input. As a result of the insignificant feature extraction, the performance in identifying stress is mediocre when very few examples are present in the training data. A method of AI called NLP converts readily available yet unstructured text that is available in various listings and databases into organized and normalized data that can be then examined or fed into ML algorithms [27].

ANOVA classifiers calculate the difference between the average variance within groups and the variance between groups. A similar variance between groups and the same mean difference between groups result in the same mean difference between groups. ANOVA focuses on the ratio of within-group variance to between-group variance. Through the use of physiological data, AI and ML have been very successful in identifying stress. For unseen anomaly cases, techniques like supervised learning and unsupervised learning have yielded 98% accuracy.

7.3 Research methodology

Figure 7.1 shows the proposed methodology for stress detection using the LSTM network.

7.3.1 Dataset

The ECG-ID dataset contributed to the PhysioNet database was used for our analysis and research [28]. Educational institutions and research institutes have widely used this database to identify and categorize mental health illnesses. Digital recordings were digitized at 500 Hz with a resolution of 12 bits over a range of 10 mV. An ECG with 12 leads is widely used to diagnose a variety of cardiac abnormalities. Several attempts have been made in the past decade to encourage 12-lead ECG classification. Several algorithms exhibit promising potential for identifying cardiac problems accurately. Nevertheless, most of these techniques have only been developed or tested on small, homogeneous, or single datasets.

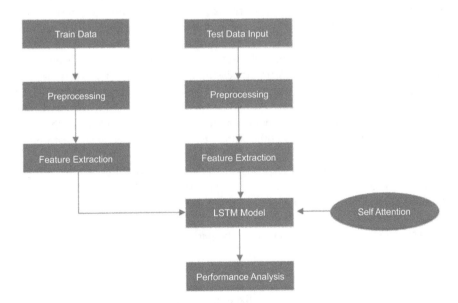

Figure 7.1   Proposed flow diagram

A variety of data sources make the PhysioNet database a valuable resource for addressing this problem. Around 300 ECG files have been collected from 90 individuals in the dataset. An ECG recording includes a text file in WFDB header format that describes the recording and patient characteristics, including diagnosis (the labels), and a binary MATLAB file containing the ECG signals. A binary file can be read using MATLAB's load function.

We obtained ECG recordings from 40+ men and 45+ women volunteers (including school/college students, friends, and coworkers of the author) between the ages of 13 and 75. We collected recordings periodically, ranging from 2 (within a day) to 20 (over 6 months). After dividing the total dataset into training and test input, 80% is used for training and 20% for testing.

7.3.2 Preprocessing the test data

The signals on a human ECG are susceptible to a variety of disturbances due to a variety of reasons. Interference from power lines, contact noise from electrodes, muscle contractions, baseline drifting, etc. are the main sources of noise. The reduction of these noise components is an essential part of ECG preprocessing. Preprocessing involves transforming raw, unstructured ECG data into a pattern that can be modeled or analyzed. In particular, it involves a set of procedures for eliminating noise and extracting relevant information. An ECG filtering technique is used to remove noise and avoid wave loss in the gathered ECG. Thus, the intended preprocessing strategy helps to remove artifacts more effectively.

A noisy ECG signal was collected and transformed into the frequency domain using a fast Fourier transform (FFT). It is possible to preserve the observed signals' peak to a greater extent by applying FFT to the ECG signals. Daubechies DB4 wavelets were used to eliminate baseline wandering from the collected ECG signal [29]. Since discrete wavelet transform (DWT) performs multi-level decomposition, this preprocessing technique can effectively capture and represent both high-frequency transients and low-frequency components of an ECG signal [22,24]. Once the approximations are subtracted from the observed signal, the noiseless signal is obtained. DWT filters have been shown to provide better performance than other filters by generating a good signal-to-noise ratio of 11.777.

As $h(n)$ presents the high-pass operational signal, $g(n)$ represents the impulse response and $x(n)$ represents the DWT signals (Figure 7.2). Low-pass and high-pass filters are used in every step of UWT to estimate the parameters of the upstream samples.

The signal's wavelet coefficient can be derived using the following equation:

$$W = W(\sigma N) + W(S) \tag{7.1}$$

where σ and N are randomly initialized noise levels, S refers to the function obtained from noisy data, and W is the wavelet transform (WT).

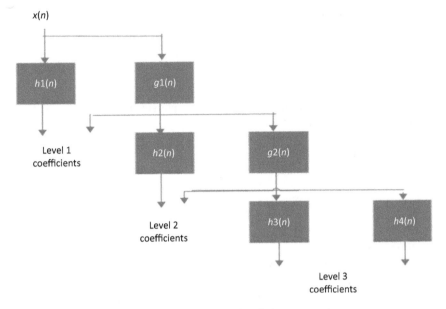

Figure 7.2 DWT multi-level decomposition

7.3.3 *Feature extraction*

The wavelet-based detector enables the extraction of time-series features. The heart rate variability (HRV), QRS complex, typical linear, statistical, nonlinear segments, and various other attributes of the ECG signal were extracted using Daubechies wavelet. The below discussed six well-known features are used for the analysis.

- **Mean**

  This measure is often called the mean since it is calculated based on the aggregate value of the input data divided by the number of records. The mean value of the signal is calculated by using (7.2).

  Assuming the input signal to be X_n, where $i = 0, 1, 2, \ldots, N$ and the mean is derived as follows:

  $$\mu_x = \frac{1}{n}\sum_{i=1}^{N} X_n \tag{7.2}$$

- **Median of the signal**

  The median of the signal is calculated by using (7.3).

  $$\text{Median value(Mm)} = \frac{1}{2}(N + 1) \tag{7.3}$$

  where N represents the total number of samples.

- **Standard deviation (SD)**

  The SD value of the signal is calculated by using (7.4).

$$\sigma = \frac{1}{N-1} \sum_{n=1}^{N} (X_n - \mu_x)^2 \tag{7.4}$$

  where μ_x is the derived mean and X_n is the input signal.

- **Root mean square (RMS value)**

  The root mean square is the square root of the average value of the squared figures (7.5). For a set of values $x_1, x_2, x_3, \ldots, x_n$, the RMS value is calculated using the following formula:

$$X_{\mathrm{rms}} = \sqrt{(x_1^2 + x_2^2 + \cdots + x_n^2)/n} \tag{7.5}$$

- **Root mean square of successive differences (RMSSD) and SD of NN (SDNN) intervals**

  Using the mean heart rate over a specific time window and the time-domain features like RMS of successive differences (RMSSD) between each heart rate and SD of normal-to-normal (SDNN), RR intervals and the stress level can be identified. Using the HRV parameter along with R values in ECG, the RMSSD is derived, which is considered one of the most effective indicators to identify the heart's parasympathetic effect.

7.3.4   LSTM architecture with SoftMax

The LSTM architecture has proved to be more effective when compared to others. There are diversified—Input, Forget, and Output gates embedded within the memory in LSTM networks. The proposed model network will enable us to store the state info for an extended period which will also eradicate fading/shattering issues. The SoftMax activation layer at the end produces probabilities for each class (non-stressful and stressful). This layer will ensure that the output values are normalized and can be interpreted as probabilities.

The LSTM is widely employed in recognizing speech and emotion as it mainly processes the temporal properties of data [1,11]. This work is mostly focused on implementing LSTM to identify the various emotions as it mainly uses the time-variant ECG signals (Figure 7.3). For categorizing emotions, a four-layered network is first developed. Secondly, by using the sentiment classifier in LSTM, the features of ECG will be retrieved along with the help of the WT.

In ML, we have many hidden layers for overall computation. In LSTM architecture, we will be using an input layer with many hidden layers and an output layer. This technology is already impended in vehicle detection and stock predictions where it was found to be very effective with 98.8% accuracy. This model is based on a stacked autocoder with a SoftMax qualifier, which will achieve 99.9% accuracy in stress detection. These models consider physiological data from humans and build data-driven models that behave according to the needs of unseen test cases.

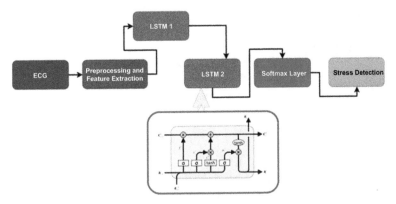

Figure 7.3   Proposed LSTM network architecture with SoftMax

7.4   Results and discussion

The baseline wanders and high-frequency disturbances were eliminated from the raw ECG signals using the Daubechies db4 wavelet-based technique as shown in Figure 7.4(a) and (b), and the filtered ECG (Figure 7.4(c)) is obtained and the hyperparameters along with HRV data are derived based on the ECG signals with RR intervals as displayed in Figure 7.4(d).

7.4.1   Performance metrics analysis

The effectuation of the designed system can be analyzed using accuracy, sensitivity, specificity, and F1-score.

The term T_p refers to the appropriately forecasted cases of loneliness, T_n stands for correctly detected normal symptoms, F_p implies to normal cases that the suggested technique misidentified as loneliness, and F_n for instances of loneliness that were wrongly classified as normal or abnormal cases.

$$\text{Accuracy} = \left(\frac{T_p + T_n}{T_p + T_n + F_n + F_p} \right) \tag{7.6}$$

$$\text{Sensitivity} = \frac{T_p}{T_p + F_n} \tag{7.7}$$

$$\text{Specificity} = \frac{T_n}{F_p + T_n} \tag{7.8}$$

$$\text{F1} - \text{score} = 2 * \frac{(\text{Recall} * \text{Precision})}{\text{Recall} + \text{Precision}} \tag{7.9}$$

where F_p represents the false positive, T_p indicates a true positive, F_n and T_n denotes false and true negatives, respectively, of the data. In our analysis, the

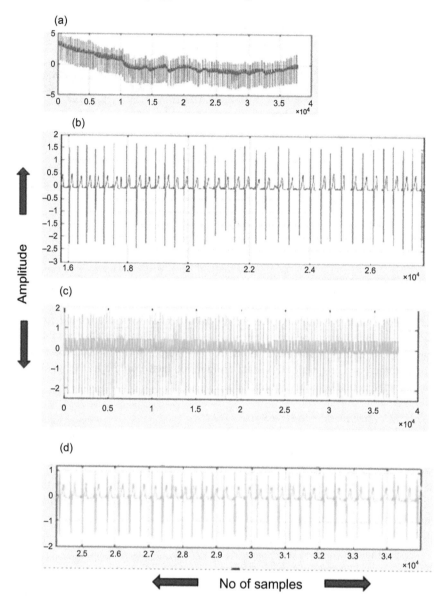

*Figure 7.4   (a) Raw ECG signal. (b) Baseline wander removed ECG signal.
(c) Filtered ECG signal. (d) ECG signal with RR intervals for validation.*

proposed model exhibited a greater accuracy (>90%) with reasonable sensitivity
(>85%) and minimal F1-score (<10%).

   The dataset obtained from PhysioNet is split into training, validation, and
testing sets. This helps in evaluating the model's performance on unseen data. In

Table 7.1   Comparison of performance metrics of various AI techniques

Classifiers	Accuracy (%)	Sensitivity (%)	Specificity	F1-score (%)
CNN	92.50	83.36	94.18	24.41
ANOVA	95.00	84.29	96.87	21.23
KNN	98.00	88.69	91.95	16.41
LSTM	98.80	92.05	89.65	14.21
LSTM with SoftMax qualifier	**99.90**	**93.44**	**98.99**	**10.72**

Note: The bold values refer to the maximum performance achieved by the proposed LSTM classifier in stress detection compared to the conventional classifiers.

the SoftMax regression model, the input layer has neurons equal to the number of extracted features, and the output layer has neurons equal to the number of stress levels. To gain a more comprehensive understanding of the model's behavior, the additional performance parameters which include Recall, F1, and precision are analyzed for the ECG signals obtained from the PhysioNet database. The derived metrics provide information about our models' performance on the dataset, their accuracy in classifying the data, and the difference between the anticipated and actual values. Greater accuracy and high sensitivity along with less F1-score concludes the model as an efficient architecture. LSTM network along with SoftMax qualifier has yielded good results when compared to the other techniques. The comparison between the sensitive parameters is depicted in Table 7.1.

7.4.2   Limitations and future scope

Models of LSTM have been extensively applied in many different fields, including stress detection. Nevertheless, LSTMs have their limitations just like any other model. They use past data extensively. LSTMs may not perform optimally if the stress patterns in the data lack a distinct temporal structure or if the stress response is poorly represented in the historical context. They may still have trouble collecting dependencies that span extremely long sequences, even though they are built to capture long-term dependencies in sequential data. The "vanishing gradient" problem refers to this situation, in which the gradients decrease as they are backpropagated through time, making it challenging for the model to learn from information from distant past periods. Also, LSTM models are computationally intensive, especially for large datasets which in turn could also limit their resource constraints.

Future developments in the fields of deep learning and affective computing, as well as continuing research, should lead to further gains in the promising field of LSTM models for stress detection. Their detection systems will be improved by combining LSTMs with additional architectures or models, such as attention processes. For increased accuracy, multi-modal LSTM models can be created to combine data from several sources. More precise and individualized stress detection may result from customizing LSTM models to account for individual variances in stress responses and patterns. Developing LSTM models for long-term monitoring and carrying out longitudinal research to comprehend how stress changes over time can yield important insights into the dynamics of stress.

7.5 Conclusion

This chapter discusses stress detection techniques and the importance of affective computing in intelligent decision-making. As part of this research, we have presented a variety of AI computing methods, including unsupervised learning, CNNs, ANOVAs, LSTMs, and LSTMs combined with SoftMax classifiers. LSTM with SoftMax is an efficient model that can be used in the analysis to identify mental stress among humans based on their ECG. It will provide maximum accuracy when it comes to analyzing test cases that have not been seen before. In this chapter, the LSTM model is endorsed as a useful tool for detecting stress at an early stage and at an accurate level. A significant correlation exists between the heart rate and QRS peaks and ECG in obtaining human values. It is imperative, however, to perform a detailed analysis to determine the various features that will enhance the classification accuracy. In conclusion, affective computing techniques will revolutionize healthcare due to their capacity to learn from data.

References

[1] Agrafoti F, Hatzinakos D, and Anderson AK (2012), ECG pattern analysis for emotion detection. *IEEE Transactions on Affective Computing*, 3(1), pp. 102–115.

[2] Singh V, Krishna NR, Norzom Bhutia T, and Singh H (2022), Effects of Virtual iRest Yoga Nidra Programme on depression, anxiety, and stress of sedentary women during the second outbreak of Covid-19. *Journal of Positive School Psychology*, 6(3), pp. 3716–3722.

[3] Maaoui C and Pruski A (2010), Emotion recognition through physiological signals for human-machine communication. In: Kordic V (ed.) *Cutting Edge Robotics 2010*. InTech, pp. 317–333 (ISBN: 978-953-307-062-9).

[4] Baghizadeh M, Maghooli K, Farokhi F, and Dabanloo NJ (2020), A new emotion detection algorithm using extracted features of the different time-series generated from ST intervals Poincaré map. *Biomedical Signal Processing and Control*, 59, p. 101902.

[5] Shahsavarani AM, Abadi EAM, and Kalkhoran MH (2015), Stress: Facts and theories through literature review. *International Journal of Medical Reviews*, 2(2), pp. 230–241.

[6] Karthikeyan P, Murugappan M, and Yaacob S (2012), A study on mental arithmetic task based human stress level classification using discrete wavelet transform. *2012 IEEE Conference on Sustainable Utilization and Development in Engineering and Technology (STUDENT)*, 6–9 October.

[7]   Picard RW, Vyzas E, and Healey J (2001), Toward machine emotional intelligence: Analysis of affective physiological state. *IEEE Transactions on Pattern Analysis and Machine Intelligence*, 23(10), pp. 1175–1191.

[8]   Karthikeyan P, Murugappan M, and Yaacob S (2012), EMG signal based human stress level classification using wavelet packet transform. *International Conference on Intelligent Robotics, Automation, and Manufacturing*, vol. 330, pp. 236–243.

[9]   Karthikeyan P, Murugappan M, and Yaacob S (2011), ECG signals based mental stress assessment using wavelet transform. *2011 IEEE International Conference on Control System, Computing and Engineering*.

[10] Gedam S and Paul S (2021), A review on mental stress detection using wearable sensors and machine learning techniques. *IEEE Access*, 9, pp. 2169–3536.

[11] Bong SZ, Murugappan M, and Yaacob S (2013), Methods and approaches on inferring human emotional stress changes through physiological signals: A review. *International Journal of Medical Engineering and Informatics*, 5(2), pp. 152–162.

[12] Quintana DS, Guastella A, Outhred T, Hickie IB, and Kemp AH (2012), Heart rate variability is associated with emotion recognition: direct evidence for a relationship between the autonomic nervous system and social cognition. *International Journal Psychophysiology*, 86(2), pp. 168–172.

[13] Zheng BS, Murugappan M, and Yaacob S (2012), Human emotional stress assessment through Heart Rate Detection in a customized protocol experiment. *2012 IEEE Symposium on Industrial Electronics and Applications*, 23–26 September.

[14] Greene S, Thapliyal H, and Caban-Holt A (2016), A survey of affective computing for stress detection: evaluating technologies in stress detection for better health. *IEEE Consumer Electronics Magazine*, 5(4), pp. 44–56.

[15] Paredes P, Sun D, and Canny J (2013), Sensor-less sensing for affective computing and stress management technology. *IEEE 2013 7th International Conference on Pervasive Computing Technologies for Healthcare and Workshops*, 5–8 May.

[16] Katsigiannis S and Ramzan N (2017), DREAMER: A database for emotion recognition through EEG and ECG signals from wireless. *IEEE Journal of Biomedical and Health Informatics*, 22(1), pp. 98–107.

[17] Balabanova I, Kostadinova S, and Georgiev G (2022), Stress recognition using sound analysis, k-NN, decision tree and artificial intelligence approach. *2021 International Conference on Biomedical Innovations and Applications (BIA)*, 2–4 June.

[18] Swarna Malika R and Ravi I (2023), Stress detection using machine learning techniques. *Journal of Emerging Technologies and Innovative Research*, 10(3), pp. 54–58.

[19] Lempert KM and Phelps EA (2014), Neuroeconomics of emotion and decision making. *Neuroeconomics, Second Edition*, pp. 219–236.

[20] Jacobs SC, Friedman R, Parker JD, *et al.* (1994), Use of skin conductance changes during mental stress testing as an index of autonomic arousal in cardiovascular research. *American Heart Journal*, 128 (6 Part 1), pp. 1170–1177.

[21] Tuerxun W, Sa'ad Alshebly Y, Azami Sidek K, and Md Johar G (2019), Stress recognition using electroencephalogram (EEG) signal. *Journal of Physics: Conference Series, Volume 1502, International Conference on Telecommunication, Electronic and Computer Engineering*, 22–24 October 2019, Melaka, Malaysia.

[22] Pourmohammadi S and Maleki A (2020), Stress detection using ECG and EMG signals: A comprehensive study. *Computer Methods and Programs in Biomedicine*, 193, p. 105482.

[23] Zhang J, Yin H, Zhang J, Yang G, *et al.* (2022), Real-time mental stress detection using multimodality expressions with a deep learning framework. *Frontiers in Neuro Science* 16, p. 947168.

[24] Li Y and Guo K (2023), Research on the relationship between physical activity, sleep quality, psychological resilience, and social adaptation among Chinese college students: A cross-sectional study. *Frontiers in Psychology*, 14, p. 1104897.

[25] Mozos ÓM, Sandulescu V, and Fernandez JM (2017), Stress detection using wearable physiological and sociometric sensors. *International Journal of Neural Systems*, 27(2), p. 1650041.

[26] Reynolds CJ (2005), Adversarial uses of affective computing and ethical implications. *Program in Media Arts and Sciences, School of Architecture and Planning*, August 28, pp. 141–145.

[27] Kumari K and Das S (2022), Stress detection system using natural language processing and machine learning techniques. *19th International Conference on Natural Language Processing (ICON 2022): WNLPe-Health 2022*, December 15–18, IIT Delhi, India.

[28] Patel V and Shah A (2021), Digital multiband filter design with power spectrum analysis for Electrocardiogram signals. *2021 International Conference on Recent Trends on Electronics, Information, Communication & Technology (RTEICT)*, 27–28 August.

[29] Bagirathan A, Selvaraj J, Gurusamy A, and Das H (2021), Recognition of positive and negative valence states in children with autism spectrum disorder (ASD) using discrete wavelet transform (DWT) analysis of electrocardiogram signals (ECG). *Journal of Ambient Intelligence and Humanized Computing*, 12, pp. 405–416.

Chapter 8

Explainable deep learning models for emotion recognition using facial images

B. Vinoth Kumar[1], J. Adlene Anusha[2], K.S. Naveena[1], S. Vats[1], M. Pravaagini[1], V. Aishwarya[1], M. Naren Siddharth[2] and B.R. Arulkumara[2]

Emotions significantly shape human behavior, requiring accurate recognition. However, the lack of transparency in deep neural networks hampers interpretability. This study employs explainable deep learning to improve the transparency of emotion recognition models. While convolutional neural network (CNN) and ResNeXt excel in emotion recognition, deciphering their decision processes remains challenging due to their black-box nature. The study advocates for integrating explainable techniques to unravel the intricacies of these models. Implementing CNN and ResNeXt, along with the mapper algorithm, involves four-fold validation of each layer's output. The mapper method calculates parameters, generating visualized, analyzed, and interpreted graphs for comprehensive inter-layer comparisons. This transparent approach clarifies model interpretations, emphasizing ResNeXt's efficiency in accurate emotion recognition. Pioneering transparent emotion recognition through explainable deep learning, the study leverages CNN and ResNeXt with the mapper algorithm to amplify model transparency and affirm ResNeXt's efficacy. These findings contribute to advancing interpretable deep learning applications in emotion recognition.

Keywords: Emotions; Deep neural networks; Interpretability; Explainable deep learning; CNN; ResNeXt; Mapper algorithm

8.1 Introduction

Emotional facial expressions seem to be the most important facial signals because they offer details about a person's personality, feelings, goals, and purpose. They serve as signals to others, urging them to act in specific ways and promoting social

[1]Department of Information Technology, PSG College of Technology, India
[2]Department of Computer Science and Engineering, PSG College of Technology, India

coordination. Facial expressions play an important role in identifying the emotional state of an individual. Individuals can have different reactions to the same stimuli [1]. Humans reveal their emotional states and intentions through facial expressions. It's one of the most potent, natural, and universal signals on the planet. Automatic facial expression detection has been the focus of various research due to its practical utility in friendly robotics, medical treatment, driver fatigue detection, and many other human–computer interaction systems [2]. In the fields of computer vision and machine learning (ML), many facial expression recognition (FER) approaches have been studied to encode expression information from face representations. The connection between human and machine communication places a high value on extracting and comprehending emotions. A person's emotional state can have a significant impact on problem-solving, reasoning, memory, and decision-making abilities, among other things [3]. To increase productivity and effectiveness, a computer system must be able to understand and communicate human emotions. The human reaction varies from person to person, so developing a generic model to find the emotion has its requirements.

A significant approach to comprehending human emotion is through using significant techniques such as machine learning/deep learning (ML/DL). These techniques are given importance because the extraction of features from facial images plays a vital role in comprehending human emotions. DL's ability to handle large quantities of information makes it extremely powerful when working with unstructured data [4]. However, ML models are frequently seen as black-boxes that are difficult to decipher. Model interpretability is considered essential especially when it comes to making predictions/assumptions based on the model analysis [5]. To solve the black-box problem and to make the model more understandable, explainable features are added to serve the purpose. With explainability features, we can explain what happens in our model from start to finish. It eliminates the black-box problem by making models transparent. Explainable DL techniques improve data insights by providing a more descriptive approach to algorithms as well as more information to users [5]. Humans can understand and explain DL algorithms and neural networks using explainable DL methodologies.

Previous works utilizing the FER-2013 dataset in conjunction with ML and DL methods have encountered limitations in accurately explaining the rationale behind emotion recognition in facial images. Most existing models lack transparency in their decision-making processes, failing to provide detailed insights into how emotions are recognized. Additionally, prior approaches often focus solely on model performance without thoroughly analyzing and interpreting intermediate layers, which restricts a comprehensive understanding of the model's inner workings. Thus, this chapter will implement explainable DL using two distinct models: a convolutional neural network (CNN) and ResNeXt, leveraging the FER-2013 dataset. The CNN will serve as a control for comparison against the ResNeXt model. Initially, the dataset is preprocessed, and the CNN and ResNeXt models are trained and evaluated. After which, four-fold cross-validation is carried out, and the output is obtained for each layer. Subsequently, the Mapper algorithm will be implemented, and graphs for each layer will be plotted. In essence, the graphs generated through the Mapper algorithm facilitate a comprehensive understanding of the learned representations in the CNN and ResNeXt models, aiding in the interpretation, comparison, and potential enhancement of the models' performance in facial emotion recognition.

This research progresses facial emotion recognition through CNN and ResNeXt models, ensuring both accuracy and interpretability in predictions. Rigorous preprocessing, robust training, and meticulous cross-validation uncover feature hierarchies. Mapper algorithm-produced graphical representations support visual interpretation, while comprehensive experiments offer insights into model behavior. The study concludes by suggesting enhancements for refining facial emotion recognition in image-based contexts. Finally, the focus shifts to interpreting and analyzing each layer and comparing the two models. The first section contains the introduction of the paper followed by the literature survey in Section 8.2, and Section 8.3 discusses elaborately the system that is proposed, namely CNNs, the ResNeXt model, and the mapper algorithm followed by the fourth section which deals with the experimental results and analysis of which includes the dataset collection, oversampling, performance measures, and the graph interpretation. Section 8.5 contains the conclusion and future enhancements.

8.2 Literature survey

Although there are numerous proposed approaches for emotion recognition using ML, there are certain drawbacks to ML, including data analysis, high error susceptibility, and the challenge of effectively interpreting findings provided by algorithms. The authors of [6] built an artificial intelligence (AI) system that can recognize emotion based on facial expressions. Because they offer stronger self-learning capabilities, automatic feature generation, enhanced analytics, and scalability, this research suggests a DL architecture. They developed a DL architecture based on CNNs for image emotion identification. The Facial Expression Recognition Challenge (FERC-2013) and the Japanese female facial expression dataset are used to evaluate the suggested method's performance (JAFFE) [7]. For the FERC-2013 and JAFFE datasets, the suggested method achieves 70.14% and 88.65% accuracy, respectively. In [8], a complete assessment of deep FER is presented, which includes datasets and techniques that provide insights into fundamental difficulties. They begin by describing the accessible datasets that are extensively utilized in the literature, as well as the data selection and evaluation principles that have been established for these datasets. The fundamental pipeline of a deep FER system is then discussed, including background information and implementation alternatives for each level. They look at existing deep neural networks (DNNs) and associated training methods that are constructed for FER using both static and dynamic picture sequences, and they look at their benefits and drawbacks in terms of the state of the art in deep FER.

For FER using infrared photos, a DL model named Infra-Red Facial Expression Network (IRFacExNet) has been suggested. The residual unit and the transformation unit are used here to extract notable aspects from input photos that are important to the expressions. The retrieved traits aid in precisely detecting the emotion of the subjects under examination in [9]. EmotionNet Nano, an efficient deep CNN (DCNN) built using a human–machine collaborative design approach that combines human experience with machine meticulousness and speed to develop a DNN design optimized for real-time embedded application, was proposed as architecture in [10]. In [11], CNN, a DL model, BOVW, and HOG, which are handcrafted approaches, all utilized face characteristics. Finally, for the facial expressions classification, they employ the support vector machine (SVM) classification algorithm. In [12], DL methods such as CNN and transfer learning (TL) were used to detect faces on real-time video footage and recognize happiness, fear, sadness, anger, surprise, and neutral emotions on the faces. For this investigation, they developed their dataset for six different facial emotions.

To get better results than with only one model, a database was built that included two types of annotations in the emotion identification domain: action units and valence arousal. The approach's uniqueness is further enhanced by the type of architecture utilized to forecast emotions: a categorical generative adversarial network [13]. In [14], a DCNN model was developed for developing a highly accurate FER system using the TL approach, in which a pretrained DCNN model is adopted by modifying its dense top layer(s) suited for FER, and the model is fine-tuned with facial expression data. The dense layer(s) is trained initially, followed by tuning each of the pretrained DCNN blocks one by one, resulting in a steady rise in FER accuracy. The authors of [15] experimented with students capturing their facial images during online classes to classify facial expressions into eight different emotion kinds by the FER algorithm. Using numerous datasets, including FER-2013, CK+, FERG, and JAFFE, [16] adopted a DL strategy that can focus on critical regions of the face and produces considerable improvement over earlier models.

It is still difficult to figure out what features of the model's input drive a DNN's decisions because it is a black-box model. As a result, the creation of methods and studies that allow for the explanation of a DNN's judgments has developed into an active and diverse study area. In [17], demographic parity was employed to assess fairness, as the likelihood of model judgments had to be independent of sensitive data. They also looked at how this constraint may be implemented in the various levels of DNNs for complex data, with a focus on graph and face recognition deep networks. Face affect qualities that are more evident when users engage with XAI interfaces were found, and a multitask feature embedding was created to connect facial affect signals with participants' usage of explanations [18]. Their findings imply that when people don't employ explanations well enough, the prevalence and levels of face AU1 and AU4, as well as arousal, rise. Thus, recognizing facial emotion using explainable DL models would give more interpretation and knowledge about the models.

For better overall accuracy, mathematical and algorithmic methods were implemented. These include placing markers on the subject's face, tracking them, and extracting features to classify the facial expressions. To classify various facial expressions, [19] proposed an algorithmic method of automated marker placement that uses an optical flow algorithm to extract statistical features (mean, variance, and root mean square). The results indicate that their proposed automated marker placement algorithm effectively placed eight virtual markers on each subject's face and gave a maximum mean emotion classification rate of 96.94% using the probabilistic neural network. In [20], a facial action coding system was used to place facial action units (FAUs; virtual markers) in specific locations on the subject's face. They calculated the distance between the FAU at the center of the subject face to other markers and used it as a feature for facial expression classification. These cross-validated features were used to map six different facial emotional expressions using K-nearest neighbor (KNN) and decision tree (DT) classifiers. A mean emotion classification accuracy of 98.03% and 97.21% was achieved using KNN and DT, respectively. In [21], a triangulation method for extracting a novel set of geometric features was proposed to classify six emotional expressions (sadness, anger, fear, surprise, disgust, and happiness) using computer-generated markers. The area of the triangle, the inscribed circle circumference, and the inscribed circle area of a triangle were extracted as features and were used to classify emotions using various ML algorithms with a maximum mean classification rate of 98.17%.

INFERENCES: The following observations are made based on the literature review and are considered. The first and most important is that a comparison of multiple DL models can be used to determine the most efficient way. Second, incorporating explainable DL into the model increases the model's transparency and provides an accurate perception to the user.

8.3   Materials and methods

8.3.1   Convolutional neural network

CNNs are a subset of feed-forward neural networks. It is a popular image recognition tool. The input data which is given as input is represented as a multi-dimensional array by CNN. When there is a lot of labeled data, it works well. It has proved to be effective in various applications, especially in FER. The architecture of CNN is shown in Figure 8.1.

- **Convolutional layer**: After the input of the images of facial expressions, each pixel is considered as a separate input neuron. The convolutional layer is the core of CNN, as the layer pulls diverse information like edges, textures, and patterns from the input image, using various filters (kernels). Rectified linear unit is employed as an activation function to decide on which neurons to be activated and the output of the node is calculated.
- **Pooling layer**: After a few convolutional layers, a pooling layer is introduced to reduce the spatial dimensions of feature maps. Two types of pooling are majorly carried out: max pooling and average pooling. In max pooling, the maximum value of each region is retained while others are discarded. In average pooling, the average value of all the elements in the region is calculated.

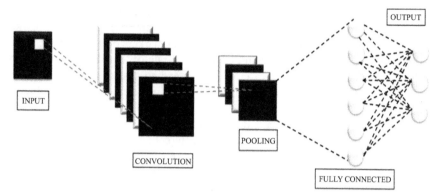

Figure 8.1   CNN architecture

- **Fully connected layers**: As it is responsible for classification, all nodes in the following layer are fully connected. Before entering the fully connected, the output of the convolution layer is normally flattened into a vector. In the output, the softmax function is utilized to provide a number between 0 and 1 based on the model's confidence. One of the seven emotions is output class. Implementing the model on the FER dataset meets both the criteria of simplicity and performance.

8.3.2   Extended residual neural network (ResNeXt)

ResNeXt is an expanded version of ResNet that is a homogenous neural network. In comparison to traditional ResNet, ResNeXt requires a smaller number of hyperparameters. It also employs cardinality, which is a third dimension in addition to ResNet's depth and width. The number of blocks is determined by the cardinality. The "split-transform merge" approach is used by ResNeXt. Instead of conducting convolutions across the whole input feature map, the block's input is projected into a sequence of lower-dimensional representations. Before merging the results, a few convolutional filters are used. The different groups in grouped convolution concentrated on different aspects of the input image, resulting in a degree of specialization among the groups. As a result, this clustered convolution technique improves accuracy. The architecture of ResNeXt is shown in Figure 8.2.

8.3.3   Overall workflow

The dataset undergoes initial preprocessing for model development, followed by training and evaluation through the plotting of accuracy and loss graphs. Subsequently, a four-fold validation is applied to each layer's output. The implementation of the mapper method is next, with ongoing visualization of graphs for each layer. Ultimately, interpretation and analysis of each layer, as well as model comparisons depicted in Figure 8.3, will be carried out.

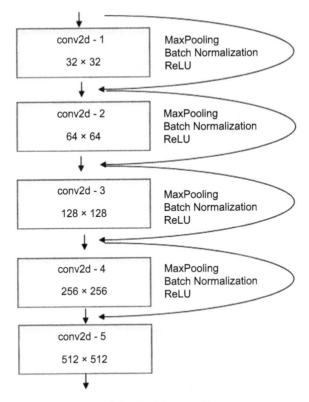

Figure 8.2   ResNext architecture

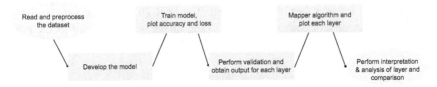

Figure 8.3   Overall workflow

8.3.4   Mapper algorithm

Our aim is to give data insights into each hidden layer of the model to give a clear picture of how the model functions in each of these layers. Explainability eradicates the black-box problem as it solves the problem of model interpretability. Model interpretability will become clear once the data and features that underpin the generated outcomes are interpreted. Explainability is all about figuring out which features contribute to the model's prediction and why they do so. Model interpretability is regarded as critical, particularly when making predictions or assumptions based on model analysis. Explainable features are added to serve the objective of solving the black-box problem and making the model more intelligible. Explainable in DL means that we can describe what happens in our model from beginning to end. Making models transparent solves the black-box problem. Explainable DL techniques boost data insights by giving consumers additional information and a more descriptive approach to algorithms. Implementation of explainable DL approaches allows humans to comprehend and explain DL algorithms and neural networks carried out through a mapper algorithm. The mapper algorithm [22] is a high-dimensional data-extraction algorithm that extracts global features. It allows for basic point cloud descriptions as simplicial complexes, abstracting away from actual distances/angles and even individual data points. The filter function, which can be any scikit-learn transformer, is defined first. It returns a subset of the data's columns. The cover is defined after the filter function has been defined. The clustering algorithm is then selected; the default clustering technique of density-based spatial clustering of applications with noise [22]. After that, the pipeline is initialized, and the mapper is plotted to obtain the results. The mapper algorithm was applied to both the models and the respective graphs, and the results for each model have been obtained.

8.4   Experimental results and analysis

8.4.1   Dataset

A training set of 28,000 labeled photographs, a validation set of 3500 labeled images, and a test set of 3500 images make up the FER-2013 dataset [23]. Each image in FER-2013 is labeled with one of seven emotions: happy, sad, angry, afraid, surprise, disgust, and neutral, with happy being the most prevalent. This provides a 24.4% baseline for guessing. The images in FER-2013 are grayscale and 48×48 pixels, with posed and unposed headshots included. The FER-2013 dataset was created by combining the results of each emotion's Google image search, as well as the synonyms for that emotion. The sample images in the dataset are shown in Figure 8.4.

8.4.2   Oversampling

Oversampling provides more measuring points, allowing averaging over a higher number of samples to improve precision [24]. Oversampling of the FER-2013 dataset gave a good increase in the effective resolution of a measurement by taking many samples. The difference in the dataset mentioned in Table 8.1 is shown as a graphical visualization before and after oversampling represented in Figures 8.5 and 8.6, respectively.

8.4.3   Performance measures

The models' performance can be assessed and examined using metrics such as precision, recall, F1-score, and accuracy. These performance measures are essential for gauging the effectiveness of a model in addressing a particular task.

4.a. Angry   4.b. Happy   4.c. Neutral   4.d. Sad

4.e. Surprise   4.f. Disgust   4.g. Fear

*Figure 8.4   Sample images from the FER-2013 dataset of seven different
emotions [23]*

Table 8.1   Dataset before and after oversampling

FER-2013 dataset	Before oversampling	After oversampling
Angry	4953	8989
Disgust	547	8989
Fear	5121	8989
Happy	8989	8989
Sad	6077	8989
Surprise	4002	8989
Neutral	6198	8989
Training set	28,709	56,630
Testing set	7178	6293
Total sample	35,887	62,923

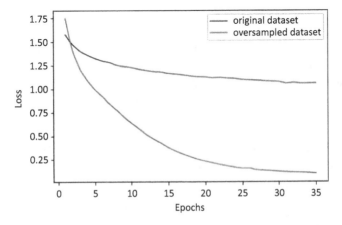

Figure 8.5   FER-2013 difference in loss due to oversampling

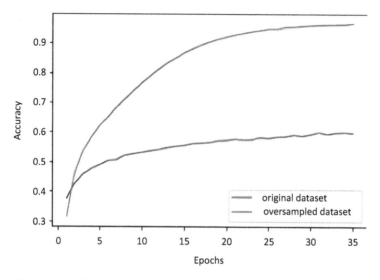

Figure 8.6   FER-2013 difference in accuracy due to oversampling

	Predicted: *NO*	Predicted: *YES*
Actual: *NO*	*TN*	*FP*
Actual: *YES*	*FN*	*TP*

Figure 8.7   Confusion matrix

- **True Positives (TP):** The number of instances that are actually positive and are predicted correctly as positive by the model.
- **True Negatives (TN):** The number of instances that are actually negative and are predicted correctly as negative by the model.
- **False Positives (FP):** The number of instances that are actually negative but are predicted incorrectly as positive by the model.
- **False Negatives (FN):** The number of instances that are actually positive but are predicted incorrectly as negative by the model.

These can be represented using the confusion matrix as shown in Figure 8.7.

- **Precision:** Precision (*P*) calculates the number of positive class predictions that are genuinely positive class predictions as shown in (8.1). It is a measure of the accuracy of the positive predictions made by a model.

$$P = \frac{TP}{(TP + FP)} \qquad (8.1)$$

- **Recall:** Recall (R) calculates the number of positive class predictions produced from the dataset's positive instances as shown in (8.2). Also called true positive (TP) or sensitivity to determine how good the model is at detecting the positives.

$$R = \frac{TP}{(TP + FN)} \qquad (8.2)$$

- **F1-score (F1):** The F1-score is the harmonic mean of precision and recall as shown in (8.3). It integrates precision and recall into a single metric to gain a better understanding of model performance.

$$F1 = \frac{2 * P * R}{(P + R)} \qquad (8.3)$$

- **Accuracy:** The accuracy is the fraction of true results among the total number of cases examined as shown in (8.4). It is an overall measure of a model's performance.

$$Accuracy = \frac{TP + TN}{(TP + FP + TN + FN)} \qquad (8.4)$$

For each model, evaluation criteria like precision, recall, F1-score, training, and testing accuracy with dataset split sizes are tabulated in Tables 8.2 and 8.3.

For further analysis, the models used in various papers listed in the literature survey along with their respective accuracies are listed in Table 8.4. The proposed

Table 8.2   Performance metrics of CNN

Evaluation metrics	Split size		
	Training size = 90% Testing size = 10%	Training size = 80% Testing size = 20%	Training size = 70% Testing size = 30%
Precision (%)	83.86	82.00	81.43
Recall (%)	82.86	80.28	80.14
F1-score (%)	83.43	81.00	80.86
Train accuracy (%)	87.60	87.31	87.85
Test accuracy (%)	**73.08**	**70.83**	**70.56**

Note: The bold denotes the overall best accuracy during the testing in CNN.

Table 8.3   Performance metrics of ResNeXt

Evaluation metrics	Split size		
	Training size = 90% **Testing size = 10%**	**Training size = 80%** **Testing size = 20%**	**Training size = 70%** **Testing size = 30%**
Precision (%)	79.43	78.28	75.14
Recall (%)	77.57	76.43	73.14
F1 Score (%)	78.14	76.85	73.85
Train Accuracy (%)	91.67	92.69	92.81
Test Accuracy (%)	**78.04**	**76.68**	**73.81**

Note: The bold denotes the overall best accuracy during the testing in ResNeXt.

Table 8.4   Results of existing works available in the literature

Reference	Paper title	Model used	Accuracy (in %)
[6]	Facial emotion detection using deep learning	A deep learning architecture based on CNN	70.14
[8]	A deep learning model for classifying human facial expressions from infrared thermal images	A deep learning model named IRFacExNet	88.43
[13]	Facial emotion recognition using transfer learning in the deep CNN	A deep CNN model using the TL approach	96.51
[15]	Emotion recognition from facial expression using deep convolutional neural network	A deep CNN approach	92.81
[18]	Optimal geometrical set for automated marker placement to virtualize real-time facial emotions	Probabilistic neural network	96.94
[19]	Facial expression classification using KNN and DT classifiers	KNN and DT classifiers	98.03
[20]	Facial geometric feature extraction-based emotional expression classification using ML algorithms	Random forest (RF) classifier	98.17

model has provided an accuracy of 73.04% using CNN and 78.04% using ResNeXt models. Although the results of the models listed in the literature survey have higher accuracies, using ResNext and integrating explainability into the proposed model has major advantages as it eliminates the black-box problem. When issues arise or errors occur, it can be challenging to debug black-box models due to their complex internal structures. Understanding and fixing errors may require extensive expertise and resources. Model interpretability is considered essential, especially when it comes to making predictions/assumptions based on the model analysis. Explainable DL techniques improve data insights by providing a more descriptive approach to algorithms as well as more information to users. Humans can understand and explain deep learning algorithms and neural networks using explainable deep learning methodologies. Therefore, it is more transparent, and the overall explainability of the model and working is better than the counterparts.

8.4.4   Graph interpretation

Graphs have been obtained in each layer for better interpretation and information on how DNNs operate using the mapper algorithm for emotion identification by using facial photos from the FER-2013 dataset, which gave a good accuracy in the CNN model in contrast to the ResNext model.

8.4.4.1   CNN

Each node in the graph represents different image data values, with these data values indicated by the concentration of red color. An increase in concentration signifies that more image values lie in the same region. Initially, in the input stage (Figure 8.8(a)), nodes are equally scattered. After passing through the first convolutional layer (Figure 8.8(b)), nodes begin forming connections, and clusters become visible. Consequently, each node starts interconnecting. In Figure 8.8(c), nodes gather in the left bottom corner of the graph, and toward the left corner, nodes start forming clusters. This implies that they have been mapped, indicating that nodes now recognize the emotion they belong to. Moving to the next graph, Figure 8.8(d), which represents MaxPooling, similarities with CNN and ResNeXt are apparent, but ResNeXt shows a more concentrated motion. All nodes have now formed a proper cluster toward the bottom left in Figure 8.8(e), which is the flatten layer. In the first dense layer (Figure 8.8(f)), the mapping is formed and linearly connected. In the second dense layer (Figure 8.8(g)), more nodes are observed to be connected.

8.4.4.2   ResNeXt

Each of the nodes in the graph represents different image data values, where these data values correspond to the concentration of the blue color. An increase in concentration signifies that more image values lie in the same region. Initially, in the input stage (Figure 8.9(a)), nodes are equally scattered. After passing through the first convolutional layer (Figure 8.9(b)), the nodes remain scattered as they struggle to determine which cluster they belong to. In Figure 8.9(c), the nodes start interconnecting, forming a mapped representation that can reveal the corresponding emotion. Moving to the next graph, Figure 8.9(d), which illustrates global average pooling, nodes gather in the left bottom corner and toward the left corner, forming clusters. In Figure 8.9(e), the flatten layer, the

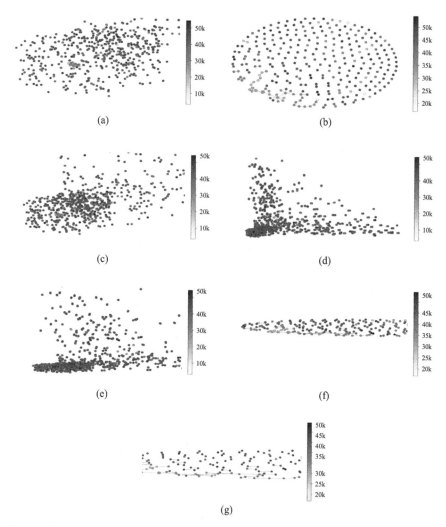

Figure 8.8   (a) Input stage graph. (b) First convolutional stage graph.
(c) Intermediate stage graph. (d) Pooling graph. (e) Flatten graph.
(f) Dense 200 stage graph. (g) Dense 7 stage graph.

concentration of nodes in the left bottom increases. In the first dense layer
(Figure 8.9(f)), a mapping is observed as nodes begin to interconnect. Finally, in
the second dense layer (Figure 8.9(g)), all nodes are strongly interconnected,
creating a fully connected and clustered node structure.

The comparison of the graphs of the CNN and ResNeXt models given in
Table 8.5 shows the gradient progressive difference between the evolution of
the stages in the model. While they start with the same input graph, the

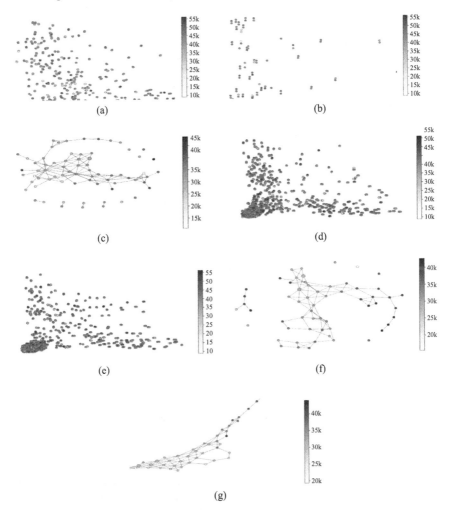

*Figure 8.9   (a) Input stage graph. (b) First convolutional stage graph.
(c) Intermediate stage graph. (d) Pooling graph. (e) Flattening graph.
(f) Dense 200 stage graph. (g) Dense 7 stage graph.*

first convolutional graph of ResNeXt shows more connectivity than the CNN
model. The consecutive neural network stages of ResNeXt however show
more scattered but interconnected plots than their CNN counterparts. As a
future scope, investigating the preservation of topological signatures as a metric
for evaluating DNN accuracy during the learning process from data holds
promise.

Table 8.5   Graph interpretation comparison

Stages	Models	
	CNN	**ResNeXt**
Input stage	After preprocessing, nodes are being scattered	After preprocessing, nodes are being scattered
First convolutional stage	Few connections are being formed	Nodes are still scattered
Intermediate stage	Nodes started accumulating and concentration has increased	Nodes are connected
Pooling stage	Max pooling has been performed	Global average pooling has been performed
Flattening stage	The concentration of the nodes is higher at the left bottom corner	Same as the pooling stage graph
Dense 200 stage	Few nodes are connected forming a linear structure	Nodes are strongly connected except for fewer anomalies
Dense 7 stage	Most of the nodes have been connected in the linear structure	Every node is interconnected and forms a closer structure

8.5   Conclusion

Explainable DL aids in the understanding of neural network stages by assisting in the determination of which model has the best accuracy, which has been demonstrated in graph interpretation as well. The FER-2013 dataset, which is used in both models (CNN and ResNeXt), provides superior accuracy in testing the ResNeXt model, as it employs backtracking, as evidenced by the graph in the Dense 7 stage, where the nodes are strongly connected. This allows us to adjust the neural network implementation to account for and strengthen the values in the dataset's range. As a result, we can better interpret the models and determine which model is most efficient in terms of graphical representation. However, using only facial expressions to detect emotion may not be accurate at all times because it can be manipulated. Developing the model into a multimodal AI model, with multiple data sources like heart rate, and videos instead of photos, can prove to be more accurate in the real world.

References

[1]   Kumar, B.V., Jayavarshini, R., Sakthivel, N., Karthiga, A., Narmadha, R., and Saranya, M. (2022). Evaluation of deep architectures for facial emotion recognition. In: Raman, B., Murala, S., Chowdhury, A., Dhall, A., Goyal, P. (eds.).

Computer Vision and Image Processing. CVIP 2021. Communications in Computer and Information Science, vol. 1567, Cham: Springer. https://doi.org/10.1007/978-3-031-11346-8_47

[2]   Rawal, N. and Stock-Homburg, R. (2022). Facial emotion expressions in human–robot interaction: A survey. *International Journal of Social Robotics*, 14. https://doi.org/10.1007/s123c69-022-00867-0.

[3]   Jung, N., Wranke, C., Hamburger, K., and Knauff, M. (2014). How emotions affect logical reasoning: Evidence from experiments with mood-manipulated participants, spider phobics, and people with exam anxiety. *Frontiers in Psychology*, 5. https://doi.org/10.3389/fpsyg.2014.00570

[4]   Chanchal, M. and Vinoth Kumar, B. (2023). Progress in multimodal affective computing: From machine learning to deep learning. In: Kumar, B.V., Sivakumar, P., Surendiran, B., and Ding, J. (eds.). *Smart Computer Vision. EAI/Springer Innovations in Communication and Computing.* Cham: Springer. https://doi.org/10.1007/978-3-031-20541-5_6

[5]   Samek, W. (2023). Chapter 2 – Explainable deep learning: concepts, methods, and new developments. In: Benois-Pineau, J., Bourqui, R., Petkovic, D., and Quénot, G. (eds.). *Explainable Deep Learning AI*, Academic Press, pp. 7–33. ISBN 9780323960984, https://doi.org/10.1016/B978-0-32-396098-4.00008-9.

[6]   Jaiswal, A., Krishnama Raju, A., and Deb, S. (2020). Facial emotion detection using deep learning. *2020 International Conference for Emerging Technology (INCET)*. https://doi.org/10.1109/incet49848.2020.9154121

[7]   Lyons, M., Kamachi, M., and Gyoba, J. (1998). The Japanese Female Facial Expression (JAFFE) Dataset [Data set]. Zenodo. https://doi.org/10.5281/zenodo.3451524

[8]   Li, S. and Deng, W. (2022). Deep facial expression recognition: A survey. *Computer Vision and Pattern Recognition*, 13(3), 1195–1215. https://doi.org/10.1109/taffc.2020.2981446

[9]   Bhattacharyya, A., Chatterjee, S., Sen, S., Sinitca, A., Kaplun, D., and Sarkar, R. (2021). A deep learning model for classifying human facial expressions from infrared thermal images. *Sciences Report*, 11(1), article no. 20696. doi:10.1038/s41598-021-99998-z.

[10] Lee, J.R., Wang, L., and Wong, A. (2021). EmotionNet nano: An efficient deep convolutional neural network design for real-time facial expression recognition. *Frontiers in Artificial Intelligence*, 3, 609673. https://doi.org/10.3389/frai.2020.609673

[11] Srinivas, M., Saurav, S., Nayak, A., and Murukessan, A.P. (2021). Facial expression recognition using fusion of deep learning and multiple features. *Machine Learning Algorithms and Applications*, pp. 229–246, Beverly, MA: Scrivener Publishing LLC. https://doi.org/10.1002/9781119769262.ch13

[12] Dandıl, E. and Ozdemir, R. (2019). Real time facial emotion classification using deep learning. *Data Science and Applications*, 2, 13–17.

[13] Richer, V. and Kollias, D. (2019). Interpretable deep neural networks for facial expression and dimensional emotion recognition in-the-wild. arXiv. https://doi.org/10.48550/ARXIV.1910.05877

[14] Akhand, M.A., Roy, S., Siddique, N., Kamal, M.A., and Shimamura, T. (2021). Facial emotion recognition using transfer learning in the deep CNN. *Electronics*, 10(9), 1036. https://doi.org/10.3390/electronics10091036

[15] Wang, W., Xu, K., Niu, H., and Miao, X. (2020). Emotion recognition of students based on facial expressions in online education based on the perspective of computer simulation. *Complexity*, 2020, 1–9. https://doi.org/10.1155/2020/4065207

[16] Liliana, D.Y. (2019). Emotion recognition from facial expression using deep convolutional neural network. *Journal of Physics Conference Series*, 1193 (1), 012004. DOI:10.1088/1742-6596/1193/1/012004

[17] Franco, D., Navarin, N., Donini, M., Anguita, D., and Oneto, L. (2022). Deep fair models for complex data: Graphs labeling and explainable face recognition. *Neurocomputing*, 470, 318–334. https://doi.org/10.1016/j.neucom.2021.05.109

[18] Guerdan, L., Raymond, A., and Gunes, H. (2021). Toward affective XAI: Facial affect analysis for understanding explainable human-AI interactions. *2021 IEEE/CVF International Conference on Computer Vision Workshops (ICCVW)*. https://doi.org/10.1109/iccvw54120.2021.00423

[19] Maruthapillai, V. and Murugappan, M. (2016). Optimal geometrical set for automated marker placement to virtualized real-time facial emotions. *PLoS One*, 11(2), e0149003. https://doi.org/10.1371/journal.pone.0149003

[20] Murugappan, M. *et al.* (2020). Facial expression classification using KNN and decision tree classifiers. *2020 4th International Conference on Computer, Communication and Signal Processing (ICCCSP)*, Chennai, India, pp. 1–6. doi:10.1109/ICCCSP49186.2020.9315234.

[21] Murugappan, M. and Mutawa, A. (2021). Facial geometric feature extraction based emotional expression classification using machine learning algorithms. *PLoS ONE*, 16(2), e0247131. https://doi.org/10.1371/journal.pone.0247131

[22] Marin, J. (2023). Explainable deep neural networks getting qualitative insights from hidden layers. https://towardsdatascience.com/explainable-deep-neural-networks-2f40b89d4d6f. Accessed on 11-12-2023

[23] Dumitru, Goodfellow, I., Cukierski, W., and Bengio, Y. (2013). Challenges in representation learning: facial expression recognition challenge. Kaggle. https://kaggle.com/competitions/challenges-in-representation-learning-facial-expression-recognition-challenge

[24] Bej, S., Davtyan, N., Wolfien, M. *et al.* (2021). LoRAS: An oversampling approach for imbalanced datasets. *Machine Learning*, 110, 279–301. https://doi.org/10.1007/s10994-020-05913-4

Chapter 9

Converging emotion recognition with AI and IoT

*Varsha Kiran Patil[1], Vijaya Rahul Pawar[2],
Aditya Kiran Patil[3], Dhanshree Yende[1], Rutika Bankar[1],
Shriya Thorat[1] and Mustafa Sameer[4]*

The combination of the Internet of Things (IoT), Big Data, and artificial intelligence (AI) has paved the way for many novel real-life situations. These converged technologies have now benefited further from the inclusion of statistics and big data analytics. Mankind is experiencing numerous applications like clever farming, business automation, and healthcare, to name a few. IoT connects the sensors, the lowest level parts in the hierarchy in remote parts of the world, to international-level service centres via consumers' fingertips, remote sensors, and servers connected to computer networks and Internet technology. Finally, this interconnection generates large volumes of huge chunks of data to be tested with statistics and to be analysed through big data analytics tools. This whole process extracts the beneficial functionality of the respective technical elements involved in the process. The new and next-generation applications and products are advertised and available with advent changes happening due to the convergence of AI, machine learning (ML), and IoT with the inclusion of emotions. This chapter addresses terminologies, implementation strategies, issues, and benefits of IoT and AI to the world. The inclusion of AI, and ML with emotions, and the results of the convergence result in big data. In addition, it results in real-life situation applications of emotion detection. Thus, this chapter is related to the gigantic world of basic concepts and applications of IoT and different algorithms of AI. Furthermore, the chapter provides the directions for accelerating the advantage of affective computing with the convergence of AI and IoT, with emotion detection.

[1]Department of Electronics and Telecommunication Engineering, AISSMS Institute of Information Technology, Pune, India
[2]Department of Electronics & Telecommunication Engineering, Bharati Vidyapeeth (Deemed to be University) College of Engineering, Pune, India
[3]Department of Electronics and Communication Engineering, Birla Institute of Technology and Sciences, Goa Campus, India
[4]Department of Electronics and Communication Engineering, National Institute of Technology, Patna, India

Keywords: Machine learning (ML); Deep learning (DL); Artificial intelligence; Internet of Things (IoT); Big data; Emotion detection; Algorithms; Data analytics

9.1   Introduction

The Internet of Things (IoT) is a popular standard that offers a variety of new services for the next trend of technical novelties. Applications in the IoT domain are almost unlimited [1], empowering all-in-one combinations of the cyber and physical worlds. Standards, professional bodies, technical alliances, industrial persons, and research scholars are working on many challenges to get the full potential of the concept of interconnecting the whole world via the Internet. On the other hand, the artificial intelligence (AI) field [2] is also blooming due to the availability of large memory, good operational speed, and advanced hardware–software configuration. The developments of the IoT as well as AI [1–5] have been responsible for advanced applications and new technological trends. One of the new trends is the inclusion of emotion recognition with statistics, pattern recognition, AI, and IoT [6]. Drawing inferences for possible solutions with the convergence of these fields will be challenging because the convergence of these trends and technology will result in unstructured and structured, big data from heterogenous sources (Figure 9.1) [7].

Big data results from the convergence of the IoT, AI, statistics, pattern recognition, and emotions [7–11]. This convergence is opening a completely new world of New Age Devices. There may be manifold ways, numerous enabling technologies, business models, and applications, and social and environmental impacts of convergence of these disruptive technologies. Disruptive technologies are the enabling technologies that changed the world's traditional approach to interconnecting devices and predicting system performance. Some of the examples of disruptive technologies that contribute to convergence are as follows:

(a) IoT
(b) AI including ML and DL, and data science [11]
(c) Convergence of AI and IoT to the Internet of Artificial Intelligence Things (AIoT)

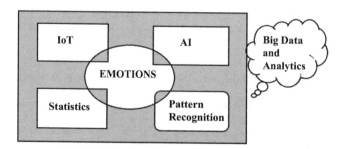

Figure 9.1   Convergence of technologies and resultant big data

(d) Big data [8]
(e) Affective computing [6]

These technologies changed device integration, prediction, and value addition in all sectors. AIoT provides combined benefits of both the IoT and AI technologies [12]. The first domain that we have surveyed is the IoT. Most of the electronic devices in homes, offices, industries, and everywhere are being connected with 5G, and the world is moving to the Internet of Everything (IoE) concept. Deliberately or unintentionally, IoT technology will be the leader in sectors of interconnecting devices for providing more advantages and comfort to humankind. The IoT implemented with voice devices [13], IoT architectural details [14], IoT developmental stages [5,12,14,15], and IoT possible applications [12,14] are the subjects of many research publications.

(a) **The journey of IoT to AIoT:** Figure 9.2 shows the stepwise convergence of IoT and AI leading to an AIoT system. The first step is sensing signals with transducers. These signals are processed by observing related patterns and waveforms. Signal conditioning is performed according to the requirements of the system. AI-based algorithms are then implemented based on the received feedback. This way, the IoT systems [16] are evolving into the Internet of Artificial Intelligence Things. The journey of IoT to AIoT is interesting and awarding in terms of enhanced applications. The IoT field evolved in 1998. The IoT field is established and ever-growing in terms of many marketplace segments, like automotive, manufacturing, strength, and healthcare. The term "Internet of Things" is framed by Kelvin Austin for the interconnection of different ubiquitously increasing (in an exponential manner) networked devices that sense and communicate securely. The IoT is useful for customer operations at anytime, anywhere. The IoT applications are useful for consumer electronics to reap deeper automation, analysis, and integration inside a system [16]. The IoT is based on the integration of various processes such as identification, recording, networking, and computation [1]. Numerous IoT applications can be grouped into different domains such as health, transportation, logistics, retail, agriculture, smart cities [15], smart metering, remote monitoring, and process automation. It is said that IoT is still in the early stages of development, despite significant advances in various research areas such as architecture, standardization, emerging technologies [16,17], and security. Useful guidelines for understanding the IoT paradigm and open issues and providing prospects for future research and development are discussed in some articles [1].

(b) **Issues of IoT and AIoT implementation and resolving:** Some of the biggest issues and challenges that the world is facing with IoT implementation are related to IoT traffic patterns and diversified traffic. As the number of devices linked to the Internet is tremendously rising, there is a requirement for newer architecture, protocols, networks, security schemes, etc. Internet protocol (IP) architecture has also been changed from IPV4 to IV6. IoT technology is having issues of safety and confidentiality, attacks, threats, vulnerabilities, and risks. Adequate authentication authorization, issues of encryption, insecure web crossing points, and bugs in software or software and programs are the issues observed for the security of cloud computing [1,14].

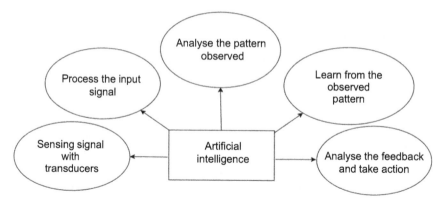

Figure 9.2   Stepwise convergence of IoT and AI leading to AIoT systems

Other challenges in IoT [15,16] development include device identification, addressability, interoperability, portability, large scale, management, energy efficiency, security, secrecy, and more. In addition, future IoT implementations must be Green IoT [1]. All these AIoT issues can be addressed with the help of AI techniques and domain knowledge.

(c) **Use of statistics for IoT and AIoT systems:** IoT structures permit advancements for the attainment of integrated solutions for sensing, networking, and robotics [14]. IoT exploits current advances in software, falling hardware prices, and cutting-edge attitudes. The development in all aspects of industrial automation and increased living standards pushed the discipline of statistics and technological know-how, to pursue the selected domain with advances in AI [18]. Since for statistics, technological know-how is broad, with some techniques applied to different disciplines, the breadth of knowledge can be extended to the welfare of society. Information technological know-how builds on skill-based expertise from technological knowledge, engineering, mathematics, statistics, and different disciplines.

(d) **Role of data science for IoT and AIoT systems:** Data science applied to domain information technological know-how is providing a completely new insight reviling with many mysteries. The data science applied to different domains provides possibilities to unlock real-time social problems like essential medical questions and urgent issues of societal importance. Designing a dashboard and graphical user interface with data science provides new inferences about operational procedures and profit margins [18].

(e) **Role of AI for IoT and AIoT systems:** The main domain under consideration for AIoT systems is AI. Alan Turing, the founding father of AI, 1950 worked on the thought of "Computer Machinery and Intelligence" [2]. AI is a group of methods that sense, recognize inputs, process, and acquire patterns with human-like intelligence. Most of the AI algorithms are about producing the methods for predictions or classification with consideration of given input. Applications of AI [18] may be related to deep learning [17], computer vision, or speech recognition [6]. AI mainly focuses on acquiring signals with reasoning with language or human sensory

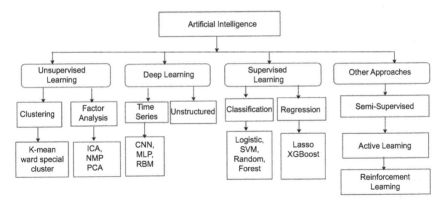

Figure 9.3   Classification of AI methods

problem-solving [2,18]. The human mimicking capabilities of AI [3] are making AI-driven systems intelligent decision-making, sensing the environment, taking suitable steps, and passing Turing tests [2] cognitively. The following chart provides an overview of AI algorithms used for solving problems in all domains. Supervised learning uses labels and is a popular method. The unsupervised has no labels but has clustering and dimensionality reduction algorithms (Figure 9.3).

(f) **AI-driven system design**

A wide range of innovations and value-added services that personalize and do interaction between users and various "things" is enabled by IoT, AI, and ML individually, or by integrations. Many research directions for performance analysis, maintenance and predictive analysis, diagnosis [8], etc. are available for varied sectors such as industries, production units, and official procedures with AI [16]. Some of the papers in the AI domain explain the algorithms implemented in applications. The converged technologies are mostly AI-driven systems. Hence the advantages of AI-based systems [18] deployed on the IoT are advantageous. These systems are classified as data preparation, simulation and modelling, deployment, and AI-based modelling (Figure 9.4).

The AI approach algorithms have progressive steps like data cleaning, model designing, simulation, and validation to impart knowledge to clouds having IoT or desktops or edge devices. The convergence of all technologies is providing research directions and avenues. The ever-expanding gigantic world of the IoT combined with AI is providing big data [9]. Big data is applied to the statistical methods and resources are utilized wisely and sufficiently. This leads to cost-cutting betterments in the products and services.

(g) **Big data:** The result of the convergence of technologies is big data. This field is important for the collection and analysis of data with the help of modern tool usage. Big data refers to datasets that might not be huge but have having big range and velocity [8,9]. This is the reason that big data may not be successfully analysed with the usage of conventional equipment and strategies. Due to the fast velocity and range of datasets or information, there is a need for a study

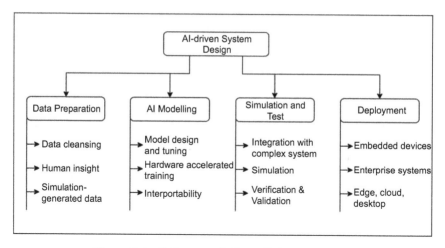

Figure 9.4   Influence of AI on AIoT applications

of the price and contents of datasets before processing and analytics. Sometimes the input data is unexpectedly changing day by day. Social community information or day-by-day transactions to consumer interactions are typical examples of day-by-day changing input data [19]. For this type of data lot of effort is required for cleaning and labelling the data. Huge information analytics and related statistics and analytics strategies with superior functionalities are part of big data [8–10]. In the literature review, it was found that huge information analytics in various domains is possible with exceptional analytics strategies [20] and equipment having special utilities.

(h) **Affective computation and integration of AI and IoT:** The above sections are about the disruptive technologies that changed traditional ways of thinking affective computing domain related to emotion detection. The world is getting benefits by integrating disruption and convergence with emotion detection, statistics, and big data analytics which otherwise would not. The predictive logic for the damage control of the systems in the use and waiting queue for scheduling algorithms is the most favourable result of the integration of the technology. Not foreseen dangers can be easily monitored and controlled by the integration [1]. The emotion detection applied to integration can be utilized for the well-being of a person's mental health [4]. The factors that are useful for emotion recognition and ways of emotion detection are to be considered while studying emotion detection.

9.2   Convergence of technologies and emotions in the education field

The involvement of emotions in the convergence of AI, ML, and big data [8–11] can be explained in the best way for the education sector. Let us consider a possible

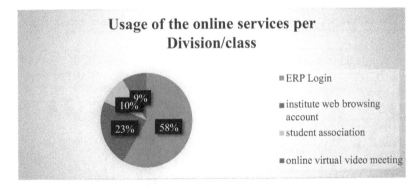

Usage of the online services per Division/class

- ERP Login
- institute web browsing account
- student association
- online virtual video meeting

9%
10%
23%
58%

Figure 9.5   Engagement of the students in online browsing

case study in which the convergence of technologies and emotions is the focal point. The education field contains information related to students, accreditation, events, placements, academic records, etc. (Figure 9.5).

Any university has several colleges. Each college has different departments, and in each department, there are several classes having a capacity of 60 students. For only one class, the following statistics and data analytics reveal how students are using online student-specific web accounts/services. The data from each department of each school or college has several such classrooms. Collecting information from many classrooms in a similar manner on the campus having different departments and organizations leads to big data. Following are some of the sample data from the learner's point of view:

- Each student in the class has an ERP account.
- Each student has an institute web browsing account.
- Each student is a member of the student association or the professional body.
- Each student has access to the learning management system (LMS).
- Each student has access to the online virtual video meeting platforms.
- Each student can access the examination test results online.
- Each student has a captive email address related to the department and organization.
- Cut-off marks for admission.
- Each student has having placement ID and there is a record of companies visited, students' internships, and placements data.
- Minimum marks and eligibility criteria for placement.
- Likes and comments on the posts shared by the stakeholders of the institute.
- Teacher's course objective mapping and course objective attainments.
- Other than listed above, there is a variety of data getting generated each day for every student, teacher, and process.

The very important point to be noted in the process of applying AIoT education field is that the students, teachers, and stockholders are live entities that get affected by emotions [4,6].

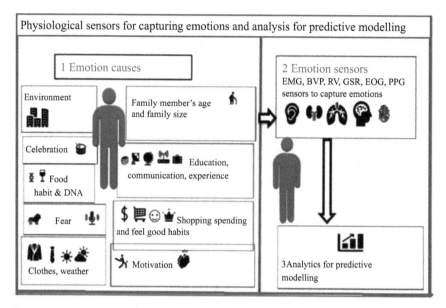

Figure 9.6   Emotions: Causes and capturing with IoT and analysis with AI tools [20,21]

The student's aptitude tests and career tests must provide faithful results because these results decide students and their future. The data engineering and analytics related to the case study of the education sector highlight and underline the amount of data, and engineering skills required for educational data analytics. Different stimuli are causing different responses in different stakeholders [4,6]. The methods of gathering emotions and the causes of the emotion arousal are listed in Figure 9.6. Three parts related to the learners emotion acquisition and processing are explained in Figure 9.6. The aspects of emotion recognition using sensors can be represented as follows:

(a) The factors causing changes in emotions.
(b) Different popular sensors and IoT
(c) Use of AI for predictive analytics/classification, etc.

9.2.1   The factors causing changes in emotions

The factors that are important for influencing emotions are listed as celebration moments, age, fear of some animals/places, DNA factors, type of clothes, environment, feel-good actors like shopping, family size, age of family members, motivation, weather, etc.

9.2.2   Different popular sensors

There are multiple ways to capture emotions. Various kinds of physiological signals [22–24] and sensor parameters can be used for emotion recognition. Invasive and noninvasive sensors are useful for acquiring a subject's body parameters for emotion recognition. Zhang *et al.* [23] have shown a dimensionality reduction method for nonlinear data. Following are some of the ways of acquiring emotions with sensors.

- Galvanic skin response (GSR)
- Respiration amplitude
- Electrocardiogram (ECG)
- Electroencephalography [25]
- Skin conductance
- Temperature skin temperature
- Heart rate
- Blood volume pulse
- Muscle impulses are some of the useful parameters to recognize emotions.

9.2.2.1 Other emotion acquisition possibilities

In the previous section, multiple methods of sensor-based emotion recognition are listed. Besides sensor-based methods, the field of emotion recognition is flourishing in diversified research areas (Figure 9.7).

(a) **DL-based emotion recognition** can be used for students' huge database of facial emotion database [18,26] mood enhancement systems [7], online

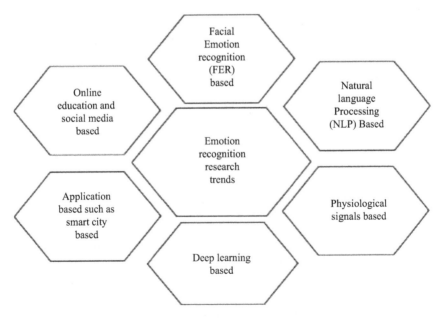

Figure 9.7 Different research trends for emotion recognition systems

engagement [27] analysis, opinion mining, etc.; mood-enhancing music recommendations [7] and suggestions for social networking [19], opinion mining, and online engagement ability measurement facilities [21,26]. All these deep learning-based examples of auto-emotion detection-based facilities can be considered as a part of "predictive emotion systems" for the entire or partial part of the educational campus. Li and Deng [28] show that DL-based facial emotion recognition (FER) systems provide better accuracy. With DL, Jung *et al.* [18] discussed DL models such as deep temporal appearance networks and deep temporal geometry networks with joint tuning. Kim *et al.* [15] mentioned another popular research direction named the emotion recognition systems have recommendations.

(b) **Physiological measures-based emotion detection systems:** Physiological measures-based emotion detection systems [22–24] use body parameters such as skin conductance, blood pressure, etc. The advantage of this approach is that people hide their emotions in facial expressions.

(c) **Natural language processing (NLP) based on emotion detection/recognition systems [20]:** To recognize emotions in text is a difficult task because the emotions are hidden and sometimes the underlying intentions are not understood easily without referring to the context. There are the following approaches: (a) classical learning-based, (b) deep learning, (c) rule-based approaches, and (d) hybrid approaches discussed in the literature [11].

(d) **Online platforms:** Some of the publicly available platforms for affective computing software packages like IntraFace have attractive features of emotion recognition. Through tweets [19], comments, posts–reposts, pictures, and speech [20] stakeholders of educational institutes are expressing their emotions. From these online platforms, emotions can be used for feedback on the activities and decisions related the campus management [21].

(e) **Smart city model:** Emotion detection with the smart city model [15] in educational campuses deals with recommendations for courses, materials, books food/cloth purchasing, health security, travelling, etc.

(f) **FER-based system:** FER-based research [24] in an educational environment can be used for viewer engagement and human activity recognition. Ayvaz *et al.* [29] mentioned that support vector machine (SVM) and *K*-nearest neighbour can be used for the FER system for online virtual classes on Skype for instant recognition of virtual learners' emotions. Dittmar *et al.* [30] directed the research of FER [25] towards automated classification of therapeutic face exercises using the Kinect, whereas the research of Han *et al.* [31] is directed towards Microsoft Kinect depth sensors-based recognition of movements for human activities [30,31]. Hossain and Muhammad [21] proposed FER-based systems to improve healthcare systems using support vector classifiers and Gaussian mixture models in smart cities. Some of the FER research directions are towards the removal of illumination and occlusions and identity bias. All facial expressions (FER) are represented by geometric features. From facial images, parameters such

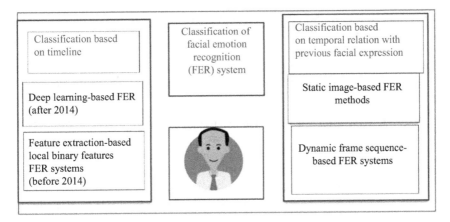

Figure 9.8   Classification trends for facial expression-based emotion detection system

as distance algorithms identify the emotions. For facial expressions from video, temporal relations with previous frames can be considered. Local binary feature-based FER popular research trend has the following general steps: (a) face representation, (b) feature extraction, and (c) classification (Figure 9.8) [18,24].

After 2014, the availability of graphical processing units and DL-based approaches accelerated, which solved memory and speed problems [18,28]. Apart from the above-mentioned methods, the speech emotion recognition (SER) approach is also yet another research direction.

9.2.3   Use of AI for predictive analytics/classification, etc.

Analytics for predictive modelling is useful for new and next-generation industrial IoT trends. But now the emotional factor is also used for predictive analysis in the fields of marketing, security, etc. Due to advancements in ML and DL [32], it is possible to extract features and perform exploratory analysis for the extraction of statistical parameters and inferences about future customers. Also, the motives of intruders can be understood before any malicious activity. Signal processing routines like calculation of entropy, Fourier transform, threshold, etc. are possible with physiological measures sensor-based systems [22,23,33]. All the descriptive and exploratory analyses along with algorithmic implementation result in the establishment of a direct or indirect relationship with emotion measurement. Predictive modelling of behaviour in terms of text in email is one of the predictive modelling examples. Dziedzicki *et al.* presented the idea of the IoT and the combination of sound signals and real-time sound-based emotion recognition are good examples of predictive modelling [33]. More details on the use of AI and predictive analytics for emotions are covered in Section 9.4.

9.3   Applications based on emotion recognition systems in the education sector

There are widespread research areas and applications of emotional areas described in the literature (Figure 9.9). Some sample representative emotion recognition applications can be seen as follows:

(a) **Virtual education in COVID-19 pandemic and again new normal solutions:**
During the COVID-19 pandemic situation in 2020, most of the countries in the world experienced a lockdown situation. At the time of lockdown and post-COVID new normal situations, emotional ups and downs were experienced. This sudden change transformed the entire education system both mentally and physically. Students and entire education shifted from offline to virtual mode. Bringing back education from online mode to virtual in the pandemic situation was a challenging task for teachers and students. The only positive thing that happened in this period was the massive use of online platforms for education, social media, and entertainment [19]. In the COVID-19 pandemic situation, the experimentation [12] during the pandemic period led to the identification of depression of the owner of the home and adding actions to improve the mood with AIoT. The impact of the new normal on virtual tools will additionally help teaching teaching-learning process with mutual agreement of both the teacher and the learner [27,29].

(b) **Recommendations systems for enhancing mood:**
Applications and systems are to be built to enhance the mood and facial expressions of music players [7]. Sharma *et al.* described such mood-enhancing recommendation systems [7].

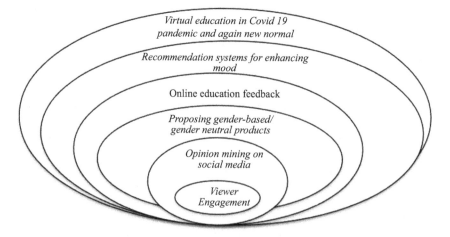

Figure 9.9   Diagrammatic representation of new and next-generation applications of emotion recognition [4]

(c) **Online education and feedback:**

Bouhlala *et al.* [5] claimed an innovative tool for improving learners' performance and learning through the use of emotion recognition. Several papers are available showing the grades or classes of online engagement with FER. Literature related to online learning linked with emotion detection is performing mood extraction of students using a neural network. The output of these experimentations is in the form of levels of engagement such as delight, boredom, frustration, neutral, and gain. Engagement of learners for online education on different online or e-learning platforms is a popular research area. Kerkeni *et al.* [6] refer to the implementation of the learner's SER system based on the Plutchiks model. Ayvaz *et al.* [29] have referred to the responses during learning. In the same paper, the emotional status is recorded by using FER systems, and learning engagement is determined based on algorithms like SVM. Dewan *et al.* [27] review various online engagement methods including FER and metrics for online engagement during online learning. Hossain and Muhammad [21] presented a combined approach of Bandlet transform, local binary pattern histograms, and an allied approach.

(d) **Opinion mining on social media:**

Extracting emotions from big data of likes, comments shares reviews, expressions, ratings, likes, and comments on social media can provide predictions about users' moods. Other research perspectives indicate comparisons of gender-wise and age-group-wise emotional responses, stability of mind, and study of different expressions for different stimuli. These directions are in turn providing directions towards response prediction of gender-based or gender-neutral products in the market, which is a new application/product trend. It is possible to make a prediction of the emotions and moods of users by considering searches and activities done by individuals on social media platforms [19]. For example, Gaind *et al.* [19] referred to the classification of text into six categories of emotions. Also, the authors have plotted time and location-wise graphs of emotions that appeared in major cities of India.

(e) **Viewer engagement:**

Online sessions analysis of facial expression gives the analysis of the understanding levels of the students. Counselling students who have difficulties in education or other issues can get better counselling if counsellors analyse the emotions with emotion analysis techniques. Based on eye movement and facial expressions, viewer engagement can be captured. The study of the emotional states of students while viewing education content and e-education is an important application.

(f) **Linkages of emotion recognition to new and next-generation issues:**

Emotion recognition research area on educational campus has several linkages to online learning platforms [20,27], affective computing, and social networking [19]. Automatic emotion recognition is a useful concept for human–computer interfaces [26]. Dziedzicki *et al.* [33] proposed various practical IoT-based emotion recognition applications (Figure 9.10).

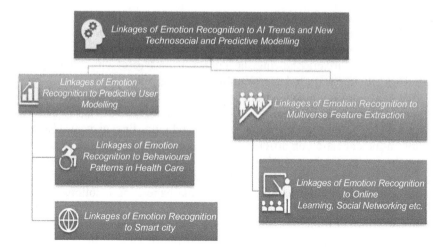

Figure 9.10   Diagrammatic representation of emotion recognition linkages

For the students having health issues, emotional states are represented in the form of different states and can be correlated to different health parameters with emotion recognition. The other application is linkages of emotion recognition to multiverse health-related feature extraction such as skin conduction via detection GSR, blood pressure and heart rate via ECG sensors, etc. [25,33]. In the literature, Marechal *et al.* [8] pointed to multiverse feature extraction [24,34] for emotion and classification. Zhang *et al.* [23] used multiple physiological inputs and did kernel principal component analysis and gradient-boosting decision tree analysis for emotion detection.

The use of augmented reality and virtual reality is also one of the major useful resources and tools for teaching the teaching-learning process. Some of the concepts in textbooks are not understandable by traditional methods of teaching. Visualization of such concepts and projecting data and concepts in an interactive and entertaining form.

9.4   Use of AI and predictive analytics for emotions

With the advent of advancements in the supporting libraries in open-source Python and OpenCV, face recognition and sensor-based emotion recognition are becoming trending [25]. The use of AI in deciding career paths and understanding interests is a very useful area of application of AI [4]. The use of AI for finding students' strengths and weaknesses blended with emotions may give students confidence. Already allotment of colleges during the college admission process using choices provided and common interest tests is done with algorithms successfully.

In the following part of the section, we will have an overview of available algorithms for affective commutative algorithms for solving possible situations in educational organizations.

(I) **Classification algorithms:** The following are commonly defined classification algorithms [3,25].

    (a) **Naive Bayes:** The Bayes theorem is considered for this algorithm where each class has a set of probabilities which is called "prior probabilities." Once the data is entered, the Naive Bayes algorithm updates prior probabilities to different classes of probability known as "posterior probability" [25].

    (b) **Decision tree:** These algorithms are based on tests representing decisions. The algorithm has nodes representing the input features. The branches show the results of the tests [25].

    (c) **Random forest:** The group of branches for which the input is split and entered to form several decision trees. The outputs of all decision trees are taken, and the average is calculated. The output of this method is more accurate than decision trees [25].

    (d) **Support vector machine:** Hyperplanes are used for the classification of data. The maximum distance between the hyperplane and the support vector is considered [25].

    (e) **K-nearest neighbours:** The data points are clustered into subgroups and then distance among centres is considered. The output is somewhat sluggish, hence is called lazy learning algorithm [25].

(II) **Regression algorithm:** Forecasting output values according to the input data points entered into the learning system. Lasso regression, logistic regression, multivariate regression, and multiple regression algorithms are the regression algorithms that are generally used in AI [3,17]. But the very simple algorithm that is popular is linear regression.

    (a) **Linear regression:** A simple algorithm, for linear relationships, or linear separable problems is called "Linear Regression." A regression line or best-fit line is a straight line drawn between data points. The line serves the prediction of new values [25].

(III) **Clustering algorithms:** Clustering is the method of an unsupervised type. Clustering is the best way in which dissimilarities are considered for separation. As well similarities are used for clustering similar data. Commonly used Clustering algorithms are as follows:

    (a) **K-means clustering:** It is the simplest unsupervised learning algorithm. The algorithm collects similar data points and binds them into one cluster. Here centroid of each group is the point of focus. K-means has K which represents the number of clusters used for grouping the data points [25].

    (b) **Fuzzy clustering algorithm:** This approach is based on probabilistic and fuzzy logic where the points are not fixed members of one cluster.

(IV) **Expectation maximization (EA) algorithm:** The data is mapped based on Gaussian distribution. The model works with the assignment of the probabilities. The expected value of the point sample is predicted with maximization equations.

    (a) **Hierarchical clustering algorithm:** This algorithm trains the data points, observes the similarities, and then sorts the clusters hierarchically through

Table 9.1   Sample data from education campus and probable approaches

Sr. no.	Data sources in the education field	Probable approaches
1	Each student in the class has an ERP account	Data science dashboards Web-based technologies
2	Each student is capable of accessing the examination test results online	Web-based technologies and optimizing tools. Clustering [10] algorithm for
3	Each student has an institute web browsing account	department-wide professional bodies algorithm
4	Each student is a member of the student association or the professional body, tracking of range of activities and involved students	
5	Each student has access to the LMS. Then the student submits assignments for autograding	Natural language processing
6	Each student has access to the online virtual video meeting platforms	Emotion analysis using FER [4] with convolutional neural network
7	Each student has a captive email address related to the department and organization	Random forest algorithm for classification of department-wise usage of the Internet
8	Teacher's course objective mapping and course objective attainments	Web-based interfaces
9	Each student has placement and there is a record of companies visited and students' internships, placements data	
10	Prediction of cut-off marks for branch-wise admission	Support vector machine

a top-down or bottom-up approach. Table 9.1 shows sample data from the education campus and probable approaches to deal with it.

9.5   All convergence results in big data [8–11]

Let us assume that there is a campus having a group of various institutions like engineering, pharmacy, management etc. The kind of data generation goes on increasing as per the level of the hierarchy from a student class, department, or institute to a campus (Figure 9.11).

Such big educational data is characterized by subsequent problems and features.

9.5.1   Features of big data

(a) **Volume of educational data:**
The quantity of data from all the time statistics increases tremendously when the data/databases are considered along with the other sensors on the campus

| Educational data from each class | Big campus has several such classes | Educational campuses produces Big Data |

Figure 9.11   Engagement of the students in online browsing

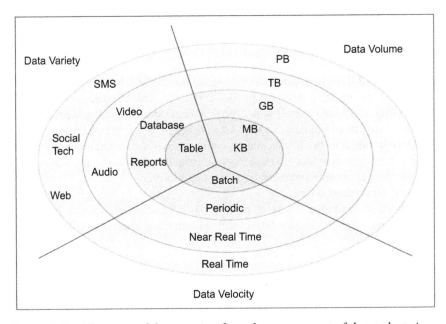

Figure 9.12   The types of data coming from the engagement of the students in online browsing [21,26]

such as CCTV capturing, smoke detectors, etc. The big data [8–11] name is related to enormous sizes or "volume" is a characteristic. If the data is big then different tools other than traditional tools are to be used for big data solutions (Figure 9.12) [8–11].

(b) **Variety of educational data:**
Student's data is heterogeneous data and data coming from different sources are structured or unstructured data. As sources are diverse, data types are dissimilar. Also, the processing capabilities of the platforms used are of a variety of kinds having special algorithms.

(c) **Velocity of educational data:**
Students exchange study materials, click pics, and exchange these pictures. The speed of educational information generation and processing meets the expected needs.

(d) **Variability of educational data:**
Student data has several kinds of inconsistencies, which bring out gaps in data. Hence the process of effectively processing and managing the data gets adversely affected. For example, assessment of the student before, during, and after the knowledge imparting may show different results.

(e) **Veracity of educational data:**
As students have mischievous minds, students' data may sometimes be filled with misleading information. The truth or accuracy of data and information assets in students' data might not be the truth [8].

(f) **The type of educational data:**

The resulting data can be categorized as follows:

• **Structured data:** If educational campus data from ERP has some fixed format and can be retrieved more easily then it is structured educational data. But if this data of several students with diversified characteristics are mixed, then it is called structured big data. Typical sizes range from a few zettabytes.

• **Unstructured data:** If educational campus data has an unknown format or structure then that data is known to be unstructured data. If unstructured data then data alignment and cleaning is a tough job [9].

• **Semi-structured data:**
If both structured and unstructured types of educational campus data are obtained, then both qualitative and quantitative analysis is done to get insights [9].

9.5.2   Data analysis

There are different tools and platforms available for big data analysis like Hadoop.

(a) **Quantitative analysis**
Numerical data is analysed and requires descriptive statistics terms like mean, median, and range. The quantitative analysis steps consist of analysis with the ML schemes, statistical analysis, etc. The relations of the variables are affecting the quantitative analysis [9].

(b) **Qualitative analysis**
In qualitative analysis [9], data is a descriptive text. The description contains information about the features in the form of shape, smell, colour, etc. The gathering of the data may have methods like educational campus data from staff, students [5], assessments interviews, videos, records of audio, and field studies. The gathered stuff provides row data. The data analysis process in qualitative analysis is based on quality whereas quantitative analysis is based on volume, weight, etc.

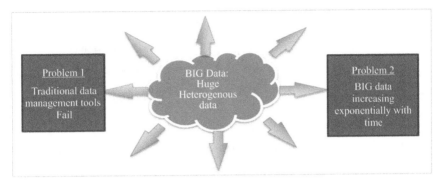

Figure 9.13 General problems in big data domain [8–11]

9.5.3 Problem due to big data [8–11]

There are two common problems in handling big data such as failure of data management tools to handle the data and generation of huge volumes of data in a short interval of time. To avoid such issues in big data, big data requires the management of structured or unstructured heterogeneous data. The data is exponentially increasing almost every second (Figure 9.13).

9.6 Use of IoT

Connecting classrooms of universities which are far apart physical distances is possible with IoT. Maintaining safety and attendance monitoring are other uses of IoT. Kim *et al.* [15] presented a detailed implementation case study for the smart city concept with minimum delay IoT devices which are connected to a virtual emotion-detectable framework, for sensing human emotions as wireless signals. Khan *et al.* [34] proposed a fusion deep learning model for an emotion-based radio frequency-based model verified for 15 participants. Big data is the result of the integration and convergence of technologies. Big data is an issue faced by the modern world on account of the increase in computing and entertainment devices in the world. The ease of clicking photographs and storing and forwarding the photos on the devices increases the possibilities of increasing the flood of information pouring into all the digital media and social media platform streams.

9.6.1 Convergence of technologies

The case study of an education system can be generalized to any field because real-time applications have multidisciplinary approaches. IoT systems together with AI have advanced big data analytics systems (Figure 9.14). Such cognitive systems are beneficial for real-time and streaming analytics. The inclusion of emotions in the convergence would increase the value of the convergence concept. The inclusion of emotion detection can recolonize the concept further. "Emotions are a complicated set of interactions among subjective and goal

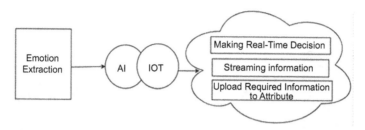

Figure 9.14   Applications based on AI, IoT, and affective computing

variables mediated through neural and hormonal systems." Furthermore, these systems can generate cognitive home automation systems together with emotion-linked AIoT-based systems [12].

This convergence has many commercial applications and initiatives. Today, huge facts have programs in pretty much convergence of AI, the IoT, and emotions in each industry like retail, healthcare, monetary services, government, customer support, etc. The manufacturing, finance, and insurance sectors have benefited from the convergence. AI and big data analytics together boon the world with the capacity to make real-time decisions [8–11]. Seize streaming information and upload required information to attributes are itself a provider of logic, and are completed while you observe [11].

Following are the prominent technological effects that occurred in the future of convergence of AIoT big data [8] and emotions.

- New job areas in cloud computing
- Industrial IoT revolution
- Increased efficiency and productivity with mood-enhancing products
- Increased data processing inferences
- Increased coverage of the healthcare system

Thus, the convergence of technology will be helpful for the betterment of humanity and the comfort of the people.

References

[1]   A. Čolaković and M. Hadžialić, "Internet of Things (IoT): A review of enabling technologies, challenges, and open research issues," *Computer Networks*, 144, 2018, 17–39. https://doi.org/10.1016/j.comnet.2018.07.017.

[2]   B. J. Copeland, "Artificial intelligence," *Encyclopaedia Britannica*, 18 March 2022, https://www.britannica.com/technology/artificial-intelligence. Accessed 23 March 2022.

[3]   H. Hassani, E. S. Silva, S. Unger, M. TajMazinani, and S. Mac Feely, "Artificial intelligence (AI) or intelligence augmentation (IA): What is the future?," *Artificial Intelligence* 1(2), 2020, 143–155. doi:10.3390/ai1020008

[4]   C. M. Tyng, H. U. Amin, M. N. M. Saad, and A. S. Malik, "The influences of emotion on learning and memory," *Centre for Intelligent Signal and Imaging Research (CISIR)*, Department of Electrical and Electronic Engineering, Universiti Teknology Petronas, Seri Iskandar, Malaysia, 24 August 2017. doi:10.3389/fpsyg.2017.01454

[5]   M. Bouhlal, K. Aarika, R. Ait Abdelouahid, S. Elfilali, and E. Benlahmar, "Emotions recognition as innovative tool for improving students' performance and learning approaches," *Procedia Computer Science* 175, 2020, pp. 597–602, ScienceDirect, *International Workshop on Artificial Intelligence & Internet of Things (A2IOT) August 9–12*, 2020.

[6]   L. Kerkeni, Y. Serrestou, M. Mbarki, and M. A. Mahjoub, "A review on speech emotion recognition: Case of pedagogical interaction in classroom," *International Conference on Advanced Technologies for Signal and Image*, 2017.

[7]   V. P. Sharma, A. S. Gaded, D. Chaudhary, S. Kumar, and S. Sharma, "Emotion-based music recommendation system," *9th International Conference on Reliability, Infocom Technologies and Optimization (Trends and Future Directions) (ICRITO)*, Noida, India, 2021, pp. 1–5. doi: 10.1109/ICRITO51393.2021.9596276.

[8]   C. Marechal, "High-performance modelling and simulation for big data applications," *Lecture Notes in Computer Science*, 2019.

[9]   S. Mishra and A. Misra, "Structured and unstructured big data analytics," *2017 International Conference on Current Trends in Computer, Electrical, Electronics and Communication (CTCIC)*, 2017, 740–746.

[10] Kefa Rabah Mara Research, Nairobi, Kenya, Convergence of AI, IoT, big data and blockchain: A review, 2018.

[11] C. Chakraborty, M. Bhattacharya, S. Pal, and S.-S. Lee, "From machine learning to deep learning: Advances of the recent data-driven paradigm shift in medicine and healthcare," *Current Research in Biotechnology*, 7, 2024. https://doi.org/10.1016/j.crbiot.2023.100164.

[12] V. K. Patil, O. Hadawale, V. R. Pawar, and M. Gijre, "Emotion linked AIoT based cognitive home automation system with sensor visual method," *2021 IEEE Pune Section International Conference (PuneCon)* 2021, 1–7. doi:10.1109/PuneCon52575.2021.9686498

[13] S. Uma, R. Eswari, R. Bhuvanya, and G. Sai Kumar, "IoT based voice/text controlled home appliances," *Procedia Computer Science* 165, 2019, 232–238.

[14] S. Parween, S. Z. Hussain, and M. A. Hussain, "A survey on issues and possible solutions of cross-layer design in Internet of Things," *International Journal of Computer Networks and Applications (IJCNA)* 8(4), 2021, 311–333. doi:10.22247/ijcna/2021/209699.

[15] H. Kim, J. Ben-Othman, S. Cho, and L. Mokdad, "A framework for IoT-enabled virtual emotion detection in advanced smart cities," *IEEE Network* 33(5), 2019, 142–148. https://dx.doi.org/10.1109/mnet.2019.1800275

[16] Tutorials point, "Internet of Things" © Copyright 2016 by Tutorials Point (I) Pvt. Ltd.

[17]   J. Liu, X. Kong, F. Xia, *et al.*, "Artificial intelligence in the 21st century," *IEEE Access*, 6, 2018, 34403–34421. doi:10.1109/ACCESS.2018.2819688.

[18]   H. Jung, S. Lee, J. Yim, S. Park, and J. Kim, "Joint fine-tuning in deep neural networks for facial expression recognition," *IEEE International Conference on Computer Vision (ICCV)*, 2015, pp. 2983–2991.

[19]   B. Gaind, V. Syal, and S. Padgalwar, "Emotion detection and analysis on social media," *Global Journal of Engineering Science and Research*, 2019, 78–89. http://www.gjesr.com/Issues%20PDF/ICRTCET-18/10.pdf

[20]   E. H. Park and V. C. Storey, "Emotion ontology studies: A framework for expressing feelings digitally and its application to sentiment analysis," *ACM Computing Surveys*, 55(9), 2023, 181. https://doi.org/10.1145/3555719.

[21]   M. S. Hossain and G. Muhammad, "An emotion recognition system for mobile applications," *IEEE Access* 5, 2017, 2281–2287. https://dx.doi.org/10.1109/access.2017.2672829

[22]   D. Ayata, Y. Yaslan, and M. E. Kamasak, "Emotion recognition from multimodal physiological signals for emotion aware healthcare systems," *Journal of Medical and Biological Engineering*, 40, 2020, 149–157.

[23]   X. Zhang, C. Xu, W. Xue, J. Hu, Y. He, and M. Gao, "Emotion recognition based on multi channel physiological signals with comprehensive nonlinear processing," *Sensors (Basel)*, 18, 2018, 3886.

[24]   A. Hassouneh, A. M. Mutawa, and M. Murugappan, "Development of a real-time emotion recognition system using facial expressions and EEG based on machine learning and deep neural network methods," *Informatics in Medicine Unlocked* 20, 100372, 2020. https://doi.org/10.1016/j.imu.2020.100372

[25]   S. Ray, "A quick review of machine learning algorithms," *2019 International Conference on Machine Learning, Big Data, Cloud and Parallel Computing (COMITCon)*, 2019, pp. 35–39. doi:10.1109/COMITCon.2019.8862451

[26]   N. Raut, "Facial emotion recognition using machine learning," *Master's Projects*, 2018, 632. https://doi.org/10.31979/etd.w5fs-s8wd

[27]   M. A. A. Dewan, M. Murshed, and F. Lin, "Engagement detection in online learning: A review," *Springer Nature Journal: Smart Learning Environments*, 6, article no. 1, 2019.

[28]   S. Li and W. Deng, "Deep facial expression recognition: A survey," *IEEE Transactions on Affective Computing* 1804, 2018, 8348–8348.

[29]   U. Ayvaz, H. Gürüler, and M. O. Devrim, "Use of facial emotion recognition in e-learning systems," *Information Technologies and Learning Tools*, 60(4), 2017, 95–104.

[30]   C. Dittmar, B. S. Olgay, J. Denzler, and H.-M. Gross (eds.), "Automated classification of therapeutic face exercises using the Kinect," *Proceedings of the 8th International Joint Conference on Computer Vision, Imaging and Computer Graphics Theory and Applications (VISAPP 2013)*, Barcelona, Spain, 2013, pp. 556–565.

[31]   J. Han, L. Shao, D. Xu, I. Member, and J. Shotton, "Enhanced computer vision with Microsoft Kinect sensor: A review," *IEEE Transactions on Cybernetics* 43(5), 2013.

[32] S. A. Oke, "A literature review on artificial intelligence," *International Journal of Information and Management Sciences* 19(4), 2008, 535–570.

[33] A. Dziedzicki, A. Kaklauskas, and V. Bucinskas, "Human emotion recognition: Review of sensors and methods," *Sensors* 20, 2020, 592. https://doi.org/10.3390/s20030592

[34] A. N. Khan, A. A. Ihalage, Y. Ma, B. Liu, Y. Liu, and Y. Hao, "Deep learning framework for subject-independent emotion detection using wireless signals," *PLoS One* 16(2), 2021, e0242946. https://doi.org/10.1371/journal.pone.0242946

Index

Printed in the USA
CPSIA information can be obtained
at www.ICGtesting.com
LVHW011819041124
795688LV00003B/264